Right-Brained Addition & Subtraction

a Forget Memorization book

Effortless learning through images, stories, hands-on activities, and patterns

by Sarah Major

www.child1st.com

Right-Brained Addition & Subtraction

© 2013 Sarah Major

ISBN: 978-1-936981-54-0

Printed in the United States of America

All rights reserved. The reproduction of any part of this book for an entire school or school system or for commercial use is strictly prohibited. No form of this work may be reproduced, transmitted, or recorded without written permission from the publisher.

Reproducing Pages from This Book
The pages in this book bearing a copyright line may be reproduced for instructional or administrative use (not for resale). To protect your book, make a photocopy of each reproducible page, then use that copy as a master for future photocopying.

To request more information regarding the copyright policy, contact:

Child1st Publications
PO Box 150226
Grand Rapids, MI 49515
800-881-0912

info@child1st.com

For other teaching and learning resources designed for visual, tactile, kinesthetic and other right-brained learners, visit www.child1st.com.

Other books and materials by this author:

Alphabet Tales
The Illustrated Book of Sounds & Their Spelling Patterns
The Easy-for-Me™ Books, sets A and B
The Easy-for-Me™ Reading Program
Kid-Friendly Math Series
Right-Brained Multiplication & Division
Writing the Visual, Kinesthetic and Auditory Alphabet
SnapWords® sight word cards and SnapLetters™ stylized alphabet
I Can Sing from 1 to 10
SnapWords® Spelling Dictionary

ABOUT THIS BOOK

This book is for children who are strongly visual, who learn all at once through pictures, are drawn to patterns, rely on body motions, and who need to understand the process behind each math problem they solve. Child1st teaching and learning resources all follow the principle of conveying learning pieces using a variety of right-brain-friendly elements. We take learning tidbits that utilize symbols (numbers and letters) and abstractions, which are left-brained, and embed them in right-brained elements to beautifully integrate the left and right hemispheres in the brain.

RIGHT-BRAINED ELEMENTS:

1- We embed symbols in *VISUALS* so that the child can take a quick look, absorb the learning piece, and store it as an image to be retrieved intact later.

2- We use *PERSONIFICATION* which is a powerful element in teaching and learning. The use of personification makes for rapid learning because the very look and personality of the character conveys the substance of the learning. For example, "8 is a snowman, makes me laugh," establishes that the symbol for 8 is really a snowman friend, and because their shapes are so similar, the child can easily use the image of the snowman to learn and remember 8.

3- We rely on *PATTERN DISCOVERY* as a way of making numbers come alive and as a means of conveying the amazing relationships between numbers. What results is number sense. Because the brain is a pattern seeking organ, it is drawn to material that follows patterns. It is my desire that through this teaching resource, many children who are overwhelmed or daunted by math might come to truly be fascinated by it instead.

4- We use *STORY* to contain the meaning of what we are teaching in math. Stories, like visuals, make learning unforgettable. They explain the "why" behind math concepts and tie everything together, creating a vehicle for meaning and for recall.

5- We use *BODY MOTION*—both gesture and whole body movement that mirrors the symbol shape or the action in the math story (such as addition or subtraction). Again, body movement is a powerful agent for learning and remembering. For many people, body motion makes recall effortless if the learning piece is directly tied to a unique motion.

6- We employ *VISUALIZATION*—a powerful tool for right-brain-dominant learners. If these learners are given time to transfer the image on the paper in front of them to their brains (prompt them to close their eyes and SEE it in their mind's eye), they will be able to retrieve that image later. If the image contains learning concepts, this is how they will remember what you want them to learn. So in this book, each time a visual is introduced, prompt the student(s) to "see" the image in their mind, eyes closed.

HOW TO USE THIS BOOK

You may approach this book in several ways, depending upon your particular needs, the level and ages of the children you are teaching, and your time constraints. If you have the time, reading the entire book is best. It is not difficult reading! Although the method of visual computing is not presented until part II (chapter 4), the chapters in part I contain essential background information. The chart on the following page shows the contents of the book globally, and Chapter 1 (p. 12) contains suggestions for use with children at a beginning or intermediate level, or for older children needing remedial assistance.

Please follow the suggested sequence of lessons on pages 4-5 if you want a guideline for what to do each day. Note that a Lesson might take more than a day, so follow your child's own pace.

Sequence of Lessons:

To familiarize yourself with the principles of teaching visual learners, please read the book through page 38.
Cut out page 7 and laminate and then display in a prominent place.
Cut out stylized numbers (page 80 and following) and arrange them on a poster board according to the directions on page 42. Display.
Note that each Lesson (below) specifies the content you will be teaching that day. Some Lessons will require multiple days to learn.
Please move at your child's pace as you move through the content of this book.

Guidelines for Teaching Lessons:
- Observe the child and follow his/her lead
- Do the activity only as long as your child needs it
- Keep on going as long as your child wants to
- Don't keep on going once his/her interest has waned. If you have to say "sit still" or "pay attention" just stop.

Lessons:

Numbers: p. 40
Review goals for the chapter on page 40. Read chapter 4 ahead of time.

☐ 1. Do Pattern Discovery on pages 42-43.

☐ 2. Learn the number song and do the body motions together. Page 45

☐ 3. Choose tactile activities from page 44 to practice forming numbers. Stay with these activities until your child is very comfortable with writing numbers. (See headings "Learn the Number Song" and "Other Kinesthetic Activities.")

☐ 4. Choose games from page 47 to help reinforce number recognition and correct number naming. Stay with these games until your child is fluent with recognizing, naming, and writing numbers.

☐ 5. Assessment: Do "Recognizing Numbers" and "Counting to Ten" on page 48.

☐ 6. Assessment, continued: Do "Ordering Numbers Sequentially" and "Supplying Missing Numbers" from page 48.

☐ 7. Assessment, continued: Do "Discovering Patterns" and "Writing Numbers" from pages 48-9.

☐ 8. Choose extension projects to do from page 49 to round out this chapter on Numbers.

Meaning of Numbers: p. 50
Review goals for the chapter on page 50. Read chapter 5 ahead of time. Make math games ahead of time or purchase the games pre-made from www.child1st.com/math/. For Lessons 9 – also consider making colorful dot cards with your child using the directions on page 52.

☐ 9. Use 3 math chips, counters, or pennies. First, show the child one chip and ask, "How many is this?" Pick up that chip and then lay down 2 chips. Ask, "How many is this?" Then move the chips a bit and ask the same thing. "Yes, there are still 2 chips." Repeat this process with 3 chips. Follow steps 1-5 on page 51. Make dot cards for 1-3 using page 52 to guide you as desired.

☐ 10. Repeat Lesson 9 but this time with 1-4 chips. Remember, we don't want the child to count dots. Better that they guess how many and then correct themselves if they guess wrong. Make dot cards for 4.

☐ 11. Repeat Lesson 9 with 1-5 chips. Please take the time your child needs to be able to say how many dots there are at a glance. (Not counting). Make dot cards for 5.

☐ 12. Repeat with 1-6 chips. Make dot cards.

☐ 13. Repeat with 1-7 chips. Make dot cards.

☐ 14. Repeat with 1-9 chips. Make dot cards.

☐ 15. Repeat with 1-10 chips. Make dot cards.

☐ 16. Play games from pages 52-53 using a set of dot cards. Repeat this Lesson until your child can instantly recognize any group of dots on any card.

☐ 17. Read the introduction found on page 53: "Visual Imprinting for Computation." Then follow the directions under "Imprinting Computations" on pages 53-54.

☐ 18. Do "Using Dot Cards for Computation Imprinting" on page 54.

☐ 19. Read the section "Visual Imprinting for Number Relationships" on pages 54-56. The cards referred to are sold on our website (www.child1st.com/math/) but you can use the reproducibles in the book or make your own as desired. Note that the Pyramid Cards are images of numbers inside a pile of 10 dots. The colored dots represent the number which added to the white dots equals 10. Play "Memory" from page 55.

☐ 20. Play "War" from page 55.

☐ 21. Play "How Many" from page 55. Continue doing this until your child is comfortable with numbers and their relationship to 10.

☐ 22. Using the section "Using the Five-Frame Cards" follow steps 1-4 on pages 55-56. Do this until the child is comfortable.

☐ 23. Do "Pattern Discovery" on pages 56-57. This make take a few days. Rule of thumb is to not force any activity. Follow the child's interest level. If you display this chart (which I recommend), your child will automatically notice other things over time.
☐ 24. Do "Becoming Familiar with Numbers to 20" on page 57.
☐ 25. Choose a Final Project from that section of page 57-58.
☐ 26. Assessment: Do the assessment as described on page 58.

Transition from Visual to Symbolic: p. 59

Review the goals for this chapter on page 59. Read all of chapter 6 and gather needed materials ahead of time.

☐ 27. Story Problem: Twitch on page 59
☐ 28. Story Problem: Pool Party on page 60
☐ 29. Story Problem: At the Playground on page 60
☐ 30. Do "Using Story Problems in the Transition to Symbols" – this might need to take more than one day. Again, keep at it as long as your child is interested and stop at the first sign of wandering attention. If you are teaching several children, you can have them act out the stories as you tell them following the suggestions under "Story Problem Activities" on page 60 at the bottom.
☐ 31. My Two Hands on page 61 – numbers 1-3 only.
☐ 32. My Two Hands on page 62 – do fours only.
☐ 33. My Two Hands – review numbers to 4.
☐ 34. My Two Hands on page 62 – do fives only.
☐ 35. My Two Hands – review numbers to 5.
☐ 36. My Two Hands for sixes.
☐ 37. Review 1-6.
☐ 38. My Two Hands for sevens.
☐ 39. Review 1-7.
☐ 40. My Two Hands for eights.
☐ 41. Review 1-8.
☐ 42. My Two Hands for nines.
☐ 43. Review 1-9.
☐ 44. My Two Hands for tens.
☐ 45. Review all numbers.
☐ 46. Do "Extension Activity" on page 63. Please refer to "Tough Spots" on page 63 if your child seems to get hung up on any of the numbers.
☐ 47. Play with encouraging your child to see My Two Hands in their imagination - see page 64.

Master Computation to 10: p. 65

Review the goals for this chapter on page 59. Read all of chapter 6 and gather needed materials ahead of time. You will need the map of Stony Brook Village and your child can have a copy of the Village – see page 69. Explore the map as you read through Chapter 7 so that you are very familiar with the concepts in this chapter before you begin to teach your child.

☐ 48. Tell the story of Stony Brook Village while you are both looking at the map. Let your child take all the time he or she wants to talk about what they notice about the map, how the numbers are placed, and any pattern they notice.
☐ 49. Review the "rules" in Story Brook Village. Today we are only going to be focused on the section "Odd and Even Houses" on page 67.
☐ 50. Today we will be doing "Comparing Streets" on page 67.
☐ 51. Do "Number Sequences" and "Houses on Streets" today. The main skills you want your child to have by this time is to understand the map, to basically know the rules of the Village..
☐ 52. From this point on, page 68 is your best friend. Each step is written out in more detail beginning on page 70-75 You will follow this sequence for each "street" starting with 3's. Today we will begin 3's. Take as long as needed to go through all 14 steps with your child before going on to the next Lesson. Please refer to page 70
☐ 53. Follow the Procedure on page 68, 71-75 for 4's.
☐ 54. Follow the Procedure on page 68, 71-75 for 5's.
☐ 55. Follow the Procedure on page 68, 71-75 for 6's.
☐ 56. Follow the Procedure on page 68, 71-75 for 7's.
☐ 57. Follow the Procedure on page 68, 71-75 for 8's.
☐ 58. Follow the Procedure on page 68, 71-75 for 9's.
☐ 59. Follow the Procedure on page 68, 71-75 for 10's.

GLOBAL VIEW of Right-Brained Addition & Subtraction

Part 1: The Framework	
Chapter	**Content**
The Children	
Chapter 1: *Consider the Child First*	**Who is targeted and why:** Identify the children who have difficulty with math and apply the right-brained method with children of all ages. Consider learning preferences and strengths contrasted with traditional methods of teaching math.
The Practice	
Chapter 2: *Use Best Teaching Practice*	**Necessary elements:** Elements of good practice for successfully teaching math to all students. How to make this process easier for everyone.
Chapter 3: *Utilize Assessments To Support Learning*	**How to assess:** Using assessments as diagnostics and as tracking tools.
Part 2: The Method	
Chapter 4: *Learn Numbers*	**Numbers:** Activities and lessons for number recognition, counting, writing numbers, and ordering numbers.
Chapter 5: *Understand the Meaning of Numbers*	**Meaning of numbers:** Activities and lessons to teach what each number means and how numbers are related.
Chapter 6: *Make the Transition from Tactile to Symbolic*	**From visual to symbolic:** Activities and lessons for real-world connections, story problems, and the action of computation.
Chapter 7: *Master Computation to Ten*	**Computing 1-10:** Activities and lessons to teach addition and subtraction with these numbers.
Appendixes	
Appendix A:	**Activities & Resources:** All materials necessary to implement the lessons in the book.
Appendix B:	**Monitoring forms:** Reproducible forms for tracking individual students and whole-class progress, geared to the content of each chapter.
Appendix C:	**Answer Keys**

TABLE OF CONTENTS

About This Book 3
Right-brained Elements 3
How to Use This Book 3
Sequence of Lessons to Follow 4-5
Global View of Right-Brained Addition & Subtraction 6
Table of Contents 7
Chart of Learning Styles 8
Map of Stony Brook 9
Introduction 11

PART I - THE FRAMEWORK 13
Chapter 1 - Consider the Child First 14
Chapter 2 - Use Best Teaching Practice 18
Chapter 3 - Utilize Assessments to Support Learning 35

PART II - THE METHOD 39
Chapter 4 - Learn Numbers 40
    Stylized Numbers to 20 in a Five Frame 41
    Number Song 45
    How to Prevent Backwards Numbers 46
Chapter 5 - Understand the Meaning of Numbers 50
Chapter 6 - Transition from Visual to Symbolic 59
    My Two Hands 61
Chapter 7 - Master Computation to Ten 65
    Procedure for Teaching Number Houses 68
    Reproducible Map of Stony Brook 69

Appendix A - Activities & Resources 79
Appendix B - Monitoring And Tracking Forms 171
Appendix C - Answer Keys 177

CHART of LEARNING STYLES

Traditional Methods are	
designed for these learners	but not for these learners!

How do I learn? (Dr. Anthony F. Gregorc)

I perceive the world

Concrete
I use my senses to take in data about the world. What I see is what is real to me.

Abstract
I visualize, intuit, imagine, read between the lines, and make connections. I pick up subtle clues.

I order the information I perceive

Sequential
I organize my thoughts in a linear, step-by-step manner. I prefer to follow a plan.

Random
I organize my thoughts in segments. I will probably skip details and even whole steps, but I will still reach the goal. I like to make up my own steps.

How do I remember? (Raymond Swassing & Walter Barbe)

Auditory
I listen to directions.
I need to hear the sounds.

Visual
I need to see it. I make visual associations, mental maps or pictures, and see patterns.

Kinesthetic
I remember well what I learn through my body. I learn best by actually doing the job.

How do I understand? (Herman Witkin)

Analytic
I am good with details, can follow steps and hear instructions, and like to finish one thing at a time.

Global
Show me the big picture! I need to see how all the parts fit in. I can hear directions after you show me the goal.

How am I smart? (Howard Gardner)

Verbal/Linguistic
I am verbal! I can speak, write, debate, and express myself well through words. IQ tests love me!

Visual/Spatial
Show me a map and I'll have it! I make vivid mental images and can use these to recall associated information. I want to see how something fits into its environment or surroundings.

Logical/Mathematical
I rely heavily on my logic and reasoning to work through problems. I am a whiz on standardized tests.

Body/Kinesthetic
I combine thinking with movement. I do well with activities that require precise motions. I learn by doing; my attention follows my movements.

© 2002 Sarah Major

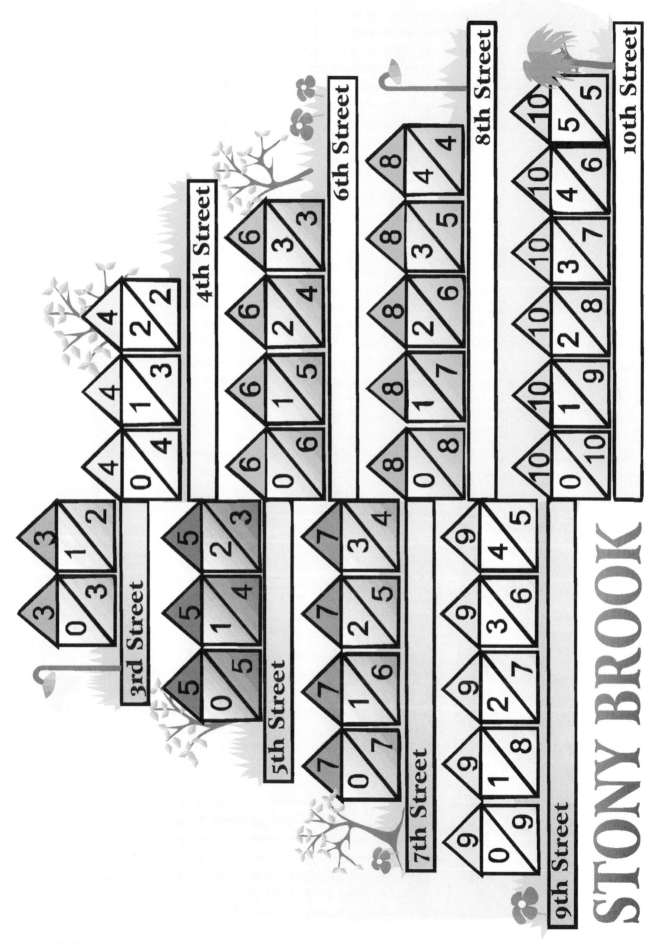

© Sarah Major 2002 | The global view of computation to ten.

INTRODUCTION - 3 Motivations For Writing This Book

Personal Experience

I am the ideal person to write a book on teaching math in a right-brained fashion. Not only am I a flaming right-brained person, I also know from personal experience what it feels like to fail arithmetic and struggle with math. What finally turned me around in my own experience with math was that during my studies in grad school, I recognized at long last that I simply could not retain what I learned when I was taught in the traditional way. As I spent valuable time discovering my own learning strengths, I took the initiative to learn and "do" math in ways that suited my own learning needs. The by-product for me was a newfound enjoyment of numbers and computation!

Personal Research

Because I am not a traditional learner, I was drawn to information about the different ways people learn, understand, remember, and are gifted. Although I'd long since learned to compensate for what was lacking in my school experiences, it was wonderful to find out there was not a deficiency in my way of seeing and remembering. I felt validated. I have learned to embrace the strengths inherent in my particular way of processing and using information.

I soon realized how often teaching practices are geared toward a few of the many learning preferences, and that children who do not happen to fall into this narrow range often suffer in their learning experiences. Frequently these learners are labeled with learning disabilities because they are unable to cope with the expectations in their school systems. Even at an early age, however, children can learn their particular learning needs and can acquire compensatory habits that will strengthen their school experience. But they must have an adult to guide them into this understanding early in their educational careers. The purpose of this book is not to teach learning styles, but where relevant I will refer to reading material that will expand on learning styles to help teachers and parents who want to better understand how to reach all their students and help them achieve success with math.

Personal Friends

A few years ago, a fourth grader, Lisa, spent the weekend at my home. When I asked her how school was going, she remarked that she just couldn't learn her "times fours." Lisa had been in a resource room for two years already and was so defeated that she rarely communicated without saying things such as, "I don't know what I'm trying to say," or "I just can't say it." She was absolutely convinced of her own inability and incompetence.

My heart went out to her, and in the few minutes we had before dinner, I decided to try to find a way to help her learn her times four table. Out of sheer blind luck, I hit on something that was based on arranging the problems in a pattern. Within 20 minutes, Lisa knew her fours. She knew them an hour later, and still knew them the next day. Two weeks later, when I saw her again, she still remembered them. What struck me during this experience with Lisa was that there was obviously nothing wrong with her brain! There was not "something missing" in her that prevented her from learning. What was missing was an approach that was compatible with the way she and many other children learn best. When she was taught in a way that dovetailed with her way of understanding and remembering, she learned rapidly and permanently.

The following spring, Lisa was tested by her school and the test results showed that she is a primarily visual learner. No learning disability emerged. Her school, however, stated that they did not have a program to accommodate her, so when she entered school that fall, she was placed in a classroom for children with learning disabilities. The glass-walled classroom, located between the two regular classrooms of the same grade level, was dubbed "The Fishbowl" by the educators in that school. Lisa had to enter that glass room every morning amidst the stares of her former friends. At night she cried. The destruction of her confidence was complete.

Since that experience with Lisa, I have worked both in introducing math for the first time to the very young and in remediation with older children, usually third graders, who were failing in school. As I approached each child determined to discover his or her particular "best learning experience" and tailor my practice to those individual needs, I experienced tremendous growth. Over time, certain global ideas emerged that were useful to all my students, regardless of their specific learning preferences.

I hope that you as parents and teachers will be filled with exuberance over the marvelous ways children's minds work, and that you will begin a lifelong habit of tailoring your teaching practice to individual children's learning needs. Although this method is excellent in remedial situations, I also hope that all teachers will begin to practice a method of teaching computation skills that will embrace all children's various ways of understanding, so that all of them can succeed within their regular classrooms.

~Sarah Major

Part I - The Framework

Chapter 1: Consider the child first

Chapter 2: Use the best teaching practice

Chapter 3: Utilize effective assessments

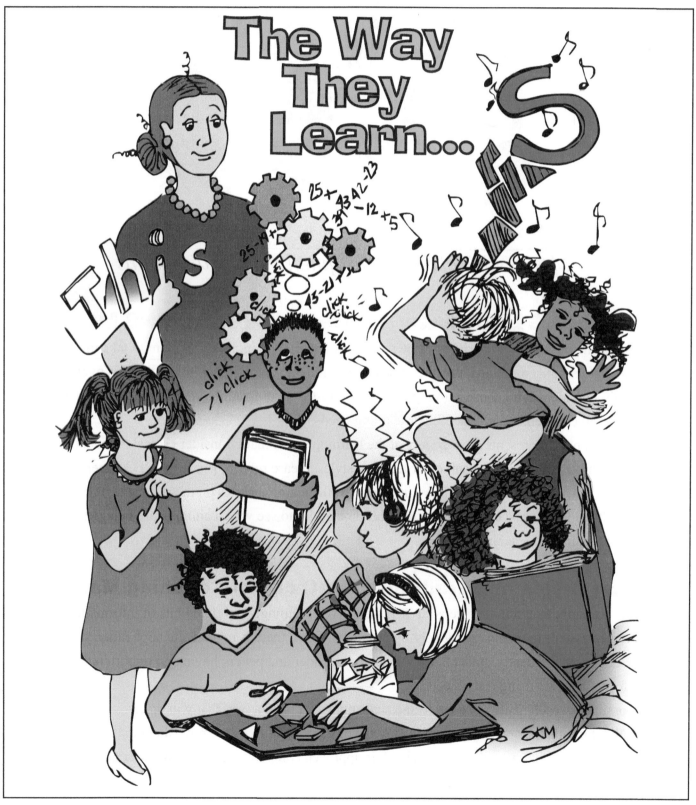

1 CONSIDER THE CHILD FIRST

Mrs. Swift's class listens quietly to her directions

CONSIDERING LEARNING STYLES WHEN TEACHING MATH

To understand the value of the right-brained method of computation, it is important to take the time to look globally at learning styles and how they relate to traditional ways of teaching math. This overview will lead naturally into identifying those learning needs that must be addressed in order for one math method to be successful for all children, regardless of their learning needs.

The chart on page 6 presents an overview of the learning styles that I consider critical to this discussion. (For in-depth information on these and other learning styles, please refer to Barbe 1985; Gardner 1993; Gregorc 1982; Tobias 1994; and Witkin 1977.) If we imagine that the learning styles on the right side of the chart represent real children in real classrooms, it will become easier to see which children are being "taught around" in traditional methods of teaching.

Learning Styles and Traditional Methods of Teaching Math

Math is normally taught in tiny steps; students are given seemingly unrelated bits of information to work with or are given steps to memorize for solving problems. Often there is no real-life application within the problems, and most frequently, students work solely with paper and pencil, having no opportunity to construct meaning for themselves using real objects.

It is possible to teach math, as concrete and sequential as it is, in a way that will reach abstract, random, visual/spatial, kinesthetic, and global students instead of continuing to try to move them over to strictly left-brained approaches. What we will do in this book is approach computation in a global, visual, kinesthetic, abstract, and random way so that no child is left out!

The Common Denominator

I have come to believe that children who are highly visual also tend to be global, somewhat random, and kinesthetic. Think about it. Visual children see a whole picture, identify smaller elements within that picture, and notice how they connect to other elements within the picture. They tend to remember parts of the picture based on where each part fits into the whole. In addition, highly visual children will move randomly through the picture (or map or pattern) and are often inclined to spatial activities that require physical skill. Visual children will prefer to see the task done as they learn it, rather than hearing it explained, and will profit from doing the problem themselves. They might not understand the process the way another student sees it, but if they are certain of the goal of the lesson, they will likely invent good steps that make sense to them and allow them to reach the goal.

Learning Disabilities?

I prefer to avoid labeling children in any way. Ever since my experience with Lisa and the fishbowl, I've been trying to learn as much as I can about learning disabilities. What are they really? Poor eyesight or hearing obviously qualify, as they could hinder learning if left undetected. Often, however, when I have inquired about the nature of a particular child's disability, the answer is something like, "She has a reading disability," or "He just cannot remember what he learns." Questions to ask before labeling a child: "Did anyone teach this child basic skills?" "Does anyone know this child's learning preference?" "Has he fallen into the habit of daydreaming?" "Has she learned to think for herself?" or "Her role in life to date has been to be cute—has someone shown her the value of problem-solving?"

Children can learn to exercise metacognition if they are guided in that direction by an observant adult. They can learn to compensate for differences they might have in learning, and they can improve their habits.

Teaching Styles

I believe it is possible to teach in such a way that every child in the group can learn without being pulled out for "special help." It would be very difficult for classroom teachers to present a lesson in seven or eight different ways in order to speak to the various intelligences, and doing so is unnecessary.

Let's adopt a teaching style that incorporates the three modalities (visual, auditory, and kinesthetic), thus encompassing three pathways to the brain. Then let's encourage the children to use their own strongest intelligences as often as possible when they do their projects. The approach becomes a flow: three pathways in, and several intelligences out.

The auditory modality is our primary means of communication. Visual and kinesthetic, though, are immensely powerful allies. Put all three together and you experience success. This right-brained approach to teaching computation is the result of the search for a way of teaching to these three modalities utilizing right-brained elements. It can be used for every child regardless of age, learning style, previous experience, or grade in school. What results is that both the "quick learners" and the "slow learners" are able to succeed. ("Quick learners" are those who can perform well in a left-brained environment.)

TEACHING TO VARIOUS AGES AND LEARNING STYLES

Here are some generalizations about how children of different ages approach learning math.

Preschool and Kindergarten

Preschool and kindergarten students need a lot of practice in a very non-pressured environment with the child in full control of the pace. Children of five or six have had relatively little experience with numbers and will need to learn to extract meaning as they are exposed to these activities. Another feature of interest is that these young children will work away, constructing meaning as they go, then suddenly, several elements will click into place for them, and they will seemingly jump ahead in their ability to manipulate and use numbers. It is important to point out the fact that children up to about age 7 are in the phase in which their right hemisphere is rapidly developing, so it only makes sense to use a right-brained approach to teaching them.

First and Second Grades

First and second graders, the challenge you will face is breaking the habit of counting on fingers. Once they understand that you are teaching them a new way of doing math that involves seeing, rather than counting on fingers, they quickly adjust to the visual-patterned method and learn very quickly.

It is normal for children of this age to master computation for a specific target number—eights, for example—in one session. By this, I mean that they can do all sums to eight and numbers subtracted from eight. I have them spend the following week practicing that target number alone, then introduce mixed practice involving other target numbers. For any non-beginner, the first challenge is to help them understand we are not using our memorizing muscle any more, but rather learning with our amazing brain camera.

Third and Fourth Grades

For nine-year-olds who have had a bad experience with math in third grade, the primary struggle will be to get them to want to try again. Many of these children have come to expect failure so much that they are stunned and seemingly cannot think. The first hurdle with this group of children is to talk them down off the cliff to which they are clinging in their anxiety. But overcoming anxiety takes many positive experiences and successes.

Traditional versus Nontraditional

For those children who are able to learn well in a regular classroom, using the Brain-Friendly Computation method helps them learn to compute more easily, prevents them from developing bad habits, and takes the tedium out of doing math. For those very visual children at the opposite end of the spectrum, this approach gives them an equal opportunity to succeed by bringing the right and left sides of the learning styles chart together somewhere in the middle. Because this first book provides a foundation for computing with multi-digit numbers, children moving into larger numbers are challenged but not daunted. Their work in *Place Value* simply refines what they have learned in this book.

CASE STUDY

During my early association with Alice, I began to believe that I would never find a strategy that would enable her to learn (that is, make good connections in her learning on her own). Nothing I did seemed to stick with her. She was stuck on 1 + 1. I was stumped. At five and a half, Alice still unconcernedly ate her way through the chocolate chips we used for computation and still did not retain anything. I

My memory of the fishbowl kept me plodding away. I learned from Alice herself just how powerfully visual she was. She is the one who inspired the "my two hands" component (explained in chapter 6), which has proved to be a simple yet powerful visual-kinesthetic learning tool.

Again, I hit upon this strategy out of desperation when I wanted to show Alice in one more way what was happening in the processes of adding and subtracting. This time, the approach clicked with her. Once we began to use "my two hands," Alice began to hum along. She took only four weeks to catch up with the rest of the class, who by this time had advanced confidently to the end of the tens.

Five and no more... One and four is five... Two and three is five.

TEACHING GUIDELINES

To recap:

1. State the goal first: "Today we are going to learn our facts for the number five," for example.

2. Provide concrete materials for the students to manipulate, establish clear but general parameters within which they will work, then let them discover the facts to five.

3. Communicate with the students about what they have discovered and guide them in drawing conclusions. This step involves pattern detection and exploration.

4. Use real-life examples of using these sums. Use stories whenever you possibly can.

5. Allow as much practice in solving problems as the students need.

6. Don't expect the students to "just remember" anything. Instead, tie every new concept to a previously learned concept, using visual and movement cues.

7. Develop a habit of teaching to all three modalities.

In the next chapter, we will discuss some practices upon which this method is based, practices that will result in a good learning experience for every learning style. These practices relate directly to the teaching of math and will ensure that this right-brained method will work for you!

2 USE BEST TEACHING PRACTICE

It was class as usual in Mrs. Swift's room!

Introducing computation in a visual way is one element of good practice, but without the right environment, it is doubtful that a new method such as this one would be effective. This chapter will touch briefly on 13 principles that will maximize the success of this method. Creating an environment conducive to good math learning might mean making some adjustments or even learning some new tricks. But the payoffs are enormous.

When I followed the guidelines in this chapter with my own students, they made tremendous gains in terms of how much they absorbed, but best of all, not a single child dreaded math or avoided it. These students have moved on into "regular" classrooms, and each of them has stunned his or her teacher by asking for more math or more difficult math. In my classroom, these children were used to asking, "Can you give me some math to do?" They were used to hearing, "What kind of problems would you like today?" If this sounds like a fantasy, let me assure you it is not.

Mrs. Swift, just moments after Alex asked for more math.

Principle 1. Engage the Emotions

Guideline: *Engaging the emotions enhances learning by creating positive biases toward math and may assist cognition.*

Rationale

Much good information has been written on the subject of the interaction of emotions and cognition (see Hart 1999; Jensen 1998; LeDoux 1996). Increasing numbers of researchers are taking the position that emotions and higher-order thinking interact (see, for example, LeDoux 1996). The bottom line for our purposes, however, is simply that the more pleasant you make the process of doing math, the more the children will like math—that formerly hated and avoided subject. Once you start to look for ways to *create positive biases toward math*, you will think of many ideas that will work well in your own classroom.

Applications

Some ideas to promote positive feelings about math:
- Remember the senses: Let a subtle scent permeate the classroom.
- Light a table lamp or two— especially for math lessons.
- Play classical music softly (no words to distract).

- Set a mood of anticipation, smile, and model positive feelings.
- Have special pencils reserved just for math use.
- Use pretty colors of paper now and then for variety.
- Let the students choose a fine-tipped marker in a favorite color for their practice.
- Choose visually appealing charts and posters depicting math patterns.
- Let the children illustrate their papers to show what is going on in the problem.
- Introduce the lesson with a short story that will engage the children and set a context for the lesson.

Principle 2. Reduce Threat

Guideline: *Negative experiences or threat during learning may detract from a student's ability to learn.*

Rationale

When the brain perceives a threat in the environment, it initiates a "fight or flight" response, the familiar adrenaline rush. Not only physical threats but also emotional or environmental threats may trigger this reflexive response, and what is perceived as threatening varies from individual to individual. In a threatening situation, one's brain is occupied with evaluating the threat and planning possible responses to it which leaves fewer mental resources for creative problem solving and learning. According to Leslie Hart (1999, 204), "Cerebral learning and threat conflict directly and completely." Hart identifies the following cognitive processes as potentially being disrupted by threat: pattern discrimination (which forms the backbone of this method), program selection (that is, "a fixed sequence for accomplishing some intended objective," such as solving a problem), the use of oral or written language, and symbol manipulation (Hart 1999, 154, 204). Obviously, if our goal is to create an environment conducive to learning, minimizing actual or potential threats in the learning environment is part of that effort.

Caleb begging to be in the YELLOW group.

Applications

Examples of possible threats include:
- Making a child do a problem on the board, then announcing that the answer is wrong

- Giving tests for which the child is not ready that result in low grades
- Ranking children in ability groups
- Forcing the whole class to progress relentlessly in lockstep despite some children's boredom and others' need to spend a little more time on a particular set of numbers
- A spirit of combativeness or competition in the group that results in the defeat of particular children, rather than a focus on whole-group cooperation and success
- An emphasis on speed and perfect papers rather than mastery, growth, and thinking processes

Principle 3. Facilitate Pattern Discovery

Guideline: *A child more readily learns material if it is embedded within a pattern*

Rationale

Brain research tells us that the human brain is a pattern-seeking organ, and we have only to watch children at work in pattern discovery to know that this is true. Patterns provide order for seemingly chaotic material and help us make sense out of seemingly random facts. Some would even argue that recognizing patterns is an innate ability (Caine, Caine, and Crowell 1999). As Leslie Hart concludes, "Learning is the extraction, from confusion, of meaningful patterns.... Even rote learning is greatly helped by detecting the patterns involved" (Hart 1999, 127). Pattern discovery also provides the global framework so important to many children. The greatest benefit of pattern discovery is that it eliminates the need for memorization of facts. Through practice, students will come to remember the pattern visually, like having a mental snapshot stored in their minds, so they can locate individual facts from within that global picture.

Applications

Suggestions for encouraging pattern discovery include the following:

- Make time for pattern discovery.
- Model your own habit of pattern discovery in everything you teach students. Ask, "How can we arrange the problems so that they form a pattern?"
- Give your students groups of facts at the same time so they can discover relationships between the problems, rather than giving them isolated problems with no apparent connection to each other (see box on page 22 for an example).
- Teach groups of facts that have something in common. (For example, present problems that have the same sum, sums that increase by one, or an addend that is the same.)
- Study and practice math facts within a pattern first, then mix them up, and finally, combine them with other, previously learned facts.

HOW TO FACILITATE PATTERN DISCOVERY

Consider the following random problem set:			
$\begin{array}{r}2\\+1\\\hline 3\end{array}$	$\begin{array}{r}4\\+2\\\hline 6\end{array}$	$\begin{array}{r}3\\+2\\\hline 5\end{array}$	$\begin{array}{r}5\\+4\\\hline 9\end{array}$
Instead, provide a set of problems with a pattern built in (each problem results in a 6):			
$\begin{array}{r}2\\+4\\\hline 6\end{array}$	$\begin{array}{r}3\\+3\\\hline 6\end{array}$	$\begin{array}{r}0\\+6\\\hline 6\end{array}$	$\begin{array}{r}1\\+5\\\hline 6\end{array}$
Ask students to figure out how to rearrange the problems that reveals more of a pattern:			
$\begin{array}{r}0\\+6\\\hline 6\end{array}$	$\begin{array}{r}1\\+5\\\hline 6\end{array}$	$\begin{array}{r}2\\+4\\\hline 6\end{array}$	$\begin{array}{r}3\\+3\\\hline 6\end{array}$
Notice in the bottom row how the numbers begin at the top, left with a zero, then 1, 2, 3, and curve down to 3, 4, 5, 6.			

Principle 4. Use a Constructivist Approach

Guideline: *Children will better understand and remember what they have worked to discover.*

Rationale

It makes sense, doesn't it, that what we work out on our own will stay in our memory longer than something someone simply tells us? The very process of working toward a specific goal will cause us to remember both what we tried that did not work and what we discovered that did work. Handing students a paper with the sums to five written on it will result in limited rote learning. In contrast, imagine the learning if we challenge students to find out which pairs of numbers will equal five, give them five counters and two bowls, and let them figure out the answers by trial and error as they share the counters between the two bowls in as many ways as they can find.

Applications

Here are some suggestions for using a constructivist approach:

• Let students figure out the rules for themselves whenever possible. You are there to stimulate the learning process, not spoon-feed answers. The process of discovery is more important than reaching a correct answer.

• Ask questions that promote "what and why" thinking, such as, "How many ways can we arrange these counters so that each new combination equals seven?" "Why do you suppose this number got bigger?" "What would happen to this problem if we changed the addend?"

• Give students real materials to represent numbers and encourage them also to draw pictures of what is happening in a problem. This is especially critical for picture thinkers.

• Provide general guidelines, such as instructing students to write down all the pairs of numbers that equal seven, then crossing off any repetitions.

• Provide specific goals for discovery.

• Design your lesson so that students will gain meaning for themselves. If you want them to discover a certain rule, choose only problems that exemplify that rule, so that in time the students will notice the rule for themselves. Make sure to alert the students that they are looking for a rule as they work.

• Instead of teaching shortcuts, give your students enough practice with similar problems so that eventually they will find shortcuts for themselves. Of course you will have to give the child your blessing as he works to discover a method that is easiest for him.

Principle 5. Give Immediate Feedback

Guideline: *Learning is enhanced by including in instruction a tool for immediate feedback.*

Rationale

A constructivist approach that utilizes pattern discovery will be greatly enhanced if some means of immediate feedback is integrated into it. Feedback is a critical ingredient in the recipe for success. The children must have a way of knowing whether the pattern they have discovered is accurate. If they find they are wrong, they must be able to make immediate revisions.

Applications

Feedback does not have to come from you! A carefully planned activity should include some form of feedback inherent in the process. Built-in feedback is a powerful teaching tool that offers students a sense of control over the learning process: if they can determine right away that their solution is not correct, they can continue their learning by searching for another answer. Here are several ways to provide instant feedback:

• Limit the number of materials given to each student. For example, if the goal is to discover pairs of numbers that equal five, give the children each five counters and two bowls, and instruct them to use all the chips each time they find a pair of numbers. They will not include the pair 1 + 2 because it does not use all their chips.

- Provide clear guidelines to direct their work. (For example, "You must use all the counters each time," or "You may not repeat the same combination of numbers.")
- Print each problem on an index card and write the answer on the back. The student can work the problem, then turn over the card to check the answer. Laminated practice sheets with dry-erase markers are a wonderful way to apply this concept with reusable materials.
- Have students work in pairs, with one partner working the problems and the other having the answers. They should take turns being the one with the answers.
- Teach kinesthetic or visual feedback methods that children can use independently, such as "my two hands" (page 61).

Principle 6. Work for Mastery

Guideline: *The goal of instruction is for every student to master the material.*

Rationale

The goal in teaching math must be mastery rather than grade collection. Every child can learn math, and if making this happen is our passion, we will do whatever it takes to achieve that goal. Success breeds more success, and once the threat of failure is removed, children can focus on learning.

An obvious reason to work for mastery by every child is that this goal most closely matches the reason we are in a classroom in the first place. Teachers are not grade collectors or classifiers of children; we are there to teach so that our students will learn. I partner with my students in this journey toward mastery.

Applications

Here are some suggestions that will help to build a community that works toward mastery:
- Tell the children the learning goals for each section. ("Today we are going to learn all our facts to ten.")
- Assess student progress frequently. (See chapter 3.)
- Provide reteaching or practice immediately after an assessment as needed to reinforce weak areas.
- Make a habit of giving brief, frequent reviews.
- Help your students to discover their learning preferences and tap into those strengths.
- Model for your students your confidence that they are able to learn the material.
- Emphasize hard work over "being smart."
- Set short-term goals in addition to the long-term goals.
- Enlist volunteer support for tracking progress and conducting additional practice with specific children.
- Use student pairs for tracking progress and for additional practice.
- Build a community spirit in your classroom, in which students help each other learn.
- Develop a habit of finding new ways to teach a concept whenever your current method is not successful for all children.

Principle 7. Achieve Mastery Through Practice

Guideline: *The brain will become fluent in those activities where it receives ample practice.*

Rationale

Practice in a nonthreatening environment is a critical element for successful learning. Rote memorization simply does not work for some children. If, after drilling with flash cards, some students still don't know their math facts, it is doubtful whether more of the same will ever produce the desired result. Give it up! Memorization is the least effective way to learn." Frequent practice is, after all, the way humans learn anything that is not genetically stamped on their brains. With sufficient practice, the process of deriving math facts becomes automatic and the child says, "That's easy!"

Applications

Some suggestions for replacing rote memorization include these:
- Do pattern discovery as discussed previously.
- Use a constructivist approach.
- Discover the method(s) that work best for each child as you ask each child, "How did you remember this?"
- Target particular elements for learning, rather than teaching a broad range of skills simultaneously. Explain the goals to the students at the beginning of the lesson.
- Use informal assessment not for the purpose of collecting grades, but in order to isolate facts not yet mastered.
- Make sure to ask each student which facts he or she has not yet mastered.
- Give practice sheets (which may be laminated for repeated practice) and ask the children to go through the page answering only those problems they recognize on sight.
- Take the time to enrich learning. Ask such questions as, "What else can we discover about this problem?" "What is it like?" and "How is it different from the one next to it?"
- Take the time to help students form additional connections for problems they answered incorrectly.
- Review missed problems using concrete materials, repeating the practice daily, if possible.
- Reassess to check for mastery.

At this point, the student will be ready to move on to the next area of mastery. Because the child is in charge of this learning process, it is safe to let the child think about what he or she needs to practice. If you let your students take the initiative in identifying what facts they need to practice, their overall progress will be more rapid than if you set the pace.

Principle 8. Provide Visual Connections

Guideline: *Find a means of visually connecting every new concept to something the student already knows.*

Rationale

Many scholars argue that most of what we remember enters the brain through the visual modality (Jensen 1994). When we make visual connections, either we are automatically reminded of something else, which is then linked in our memory with the new idea, or we make a conscious effort to form a connection that will serve as a memory prompt.

For visual learners, this avenue is essential. When visual learners hear verbal instructions, what they hear and what they are able to process are frequently two different things. They need to see what they are hearing.

Applications

Here are some suggestions for making connections as you teach:

• Ask students these questions constantly during your teaching: "What does this look like?" and "What does this remind you of?" Record students' answers.

• Discuss the various suggestions with the students and collectively agree on a specific visual cue for each concept. These cues become triggers that help students retrieve abstract facts from memory. Once the fact has been learned, children will automatically lose the need for the visual connection and recall will be automatic.

CASE STUDY

When my preschool students were having trouble remembering which number symbol corresponded to each number name, we discussed what each number looked like. The group chose known objects, the shapes of which reminded them of each number symbol. I watched and waited patiently during the early days of number learning, when the children

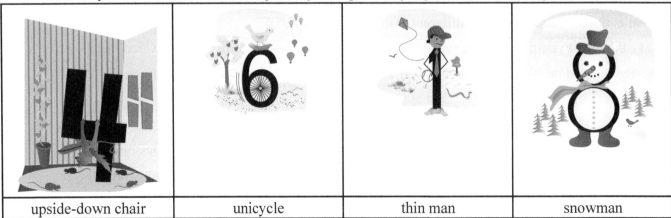

| upside-down chair | unicycle | thin man | snowman |

used the number names and associated picture names interchangeably. Finally, the day came when they dropped the picture name and retained the number name. With the extra visual step of associating a picture, learning occurred more rapidly and without stress for the children. The children were easily engaged in learning their numbers. These stylized numbers will be presented in chapter 4.

Principle 9. Set the Stage for Visual Imprinting

Guideline: *Teach your students to study a global body of material, then close their eyes and see it in their minds rather than memorizing the facts in the body of material.*

Rationale

Visual imprinting refers to a practice that is difficult to define. Imprinting just happens; it is a subconscious form of learning (what I call "learning through the back door"). Visual imprinting has occurred any time we can "see" a complete picture or a specific part of it in our mind's eye. Even though visual imprinting is elusive, we can deliberately take advantage of it in our teaching. In fact, it has become one of the most powerful tools I use in teaching. It is primarily through this means that my children learn their sight words, the meaning behind the number symbols, and their math facts to ten. Each time I ask, "How did you remember that?" and the student answers, "I saw the snowman" (see illustration below) or some other response that reveals visual imprinting, I feel the magic of visual learning all over again.

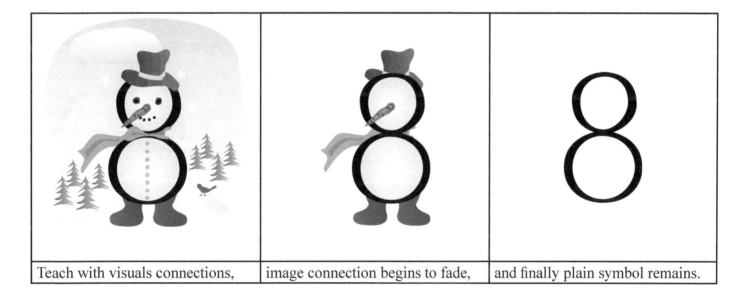

| Teach with visuals connections, | image connection begins to fade, | and finally plain symbol remains. |

Applications

To promote visual imprinting in your teaching, try these strategies:

• Use stylized materials as often as possible. For example, in helping a child remember 20, 30, 40, and so on, you could quickly make the 0 in each number into a teacup. Then, point to the first digit as you say the first part of the number name, and as you point to the 0/teacup in each number, say "tea." Let the pictures enter the mind of the child passively; do not try and actively teach the association. Remember, the child is acquiring a mental photo of the concept. This is an automatic process, not one that can be forced.

• Use such stylized materials only about three times during learning before stopping to check for recall using normal printed numbers. Isolate the facts the child still does not recall and use stylized materials with those facts a few more times.

- Recheck for knowledge using normal printed materials.
- Frequently ask various students, "How did you remember that?" This habit of yours will teach the students to think about their own learning processes.

CASE STUDIES

ETHAN:

A fascinating example of visual imprinting occurred one spring as I was helping nine-year-old Ethan learn the multiplication table for eights. We were using an approach in which I provided a five-by-four grid of answers. This global approach utilizes pattern discovery, visual mapping, and apparently also visual imprinting. After only 20 minutes, Ethan could do a sheet of mixed problems quickly and accurately without referring to the written answers and without counting. I was enthralled to see that as he worked, he would read a new problem, look at the empty grid, point to where the answer would be located, and then quickly write the answer. When I asked him what he was doing, he replied that he was finding the answers in the grid. Because Ethan was concerned that he would not be able to recall the facts in his regular classroom, I asked whether he thought his teacher would mind if he taped an empty grid to his desktop. Horrified, Ethan said, "But that would be cheating!" His answer told me just how clearly Ethan was seeing the answers to the multiplication problems on the blank grid. I suspect that a combination of visual imprinting and visual mapping on a global pattern was occurring.

BECKY:

One day Becky, who was nine, was humming through her sums for six using the "Stony Brook Village" approach, in which each sum is represented inside a house on a particular numbered street (see chapter 7). After discovering the global pattern for sixes, she was using her new knowledge to complete a practice sheet of problems. I pointed to a problem she had just finished and asked, "How did you remember this one?" She instantly replied, "Oh, that is the last house on the right." Neither drill nor memorization had taken place; learning was fast and painless. We set the stage with a story, did pattern discovery, practiced writing the pattern three times, then began some solo flights through sheets of problems. Becky was happily successful.

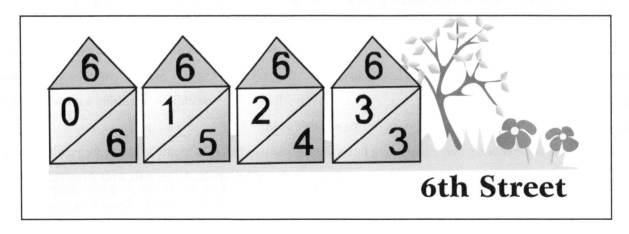

Principle 10. Make Brain-Body Connections

Guideline: *As often as possible, combine visual materials closely with body actions or movements.*

Rationale

The visual modality is a powerful means of learning, but learning is enriched and deepened when we involve the body in the process. Doing so utilizes two powerful means of remembering simultaneously.

Applications

So, to strengthen your teaching practice, try these tips:

- Make connections between visual images and physical sensations, activities, or movements. For example, you could have a child make a dot card to represent the quantity three by dipping three fingertips into colored paint and then pressing them on an index card. The combination of pressing three fingers on the card and seeing the resulting dot pattern will result in significant visual-body connections as well as provide an image for imprinting (see illustration). In this activity, the child feels three fingertips getting wet and slippery and three fingertips pressing on the card.

Beyond that, the child unconsciously gains a visual image of "threeness" not only by seeing the fingerprints on the card, but also by seeing in memory those three fingertips colored with paint. (This activity also leads naturally into the "my two hands" strategy, described on page 61.)

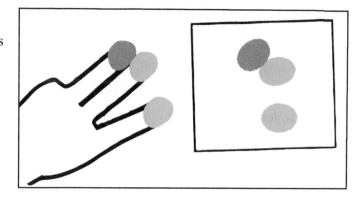

- Deliberately build triggers for recall in your students' minds: The difference in practice between doing this kind of activity versus simply counting three objects, or providing preprinted dot cards, is subtle but significant. After doing the activities, students will have images of number meanings stored in their memories. They will see three instead of counting to three. Similarly, the "my two hands" method of computation is a kinesthetic means of learning that also promotes heavy visual imprinting. Many children actually see mental images of the sums after having practiced them on their hands.

Principle 11. Make Associations and Connections

Guideline: *Teach every new concept as an outgrowth of previous learning and a connection to future learning.*

Rationale

Associations and connections are the best aids for memory and recall. When the brain is confronted with new, seemingly chaotic information, it gets busy trying to form patterns, discover connections to other knowledge, and make associations that will help it to make sense of and remember the new information. Thus, math learning can be greatly enhanced by these strategies:

- Assist the child to form associations with real events.
- Connect the steps in the learning process to one another.

Rather than teaching math in unrelated segments, we can start out purposefully with simple concepts upon which more complex skills will be built. This practice is consistent with research showing that the brain learns a basic skill and then develops more specialized versions of that initial skill (Hart 1999, 177). Start small, as you would when you pack a snowball in your hand. Then add to that first fact as you would add snow to the snowball by rolling it across the snowy yard.

One example of how this strand runs throughout this method is the use of the number five. Students start by seeing what five looks like on their own hand (five fingers), then move to the use of a fives chart, then to using multiples of five as anchor numbers (that is, as reference points against which to describe the position of other numbers), to using groups of five in learning multiplication and division. All this is done purposefully to build on a simple concept that the child learned at the very beginning.

Applications

Here are some suggestions for making associations and connections in your teaching:

- Plan lessons that integrate new information with prior learning.
- Teach information in groups of related facts rather than as isolated facts.
- Find as many connections as possible to real-world situations.
- Tie new information to other disciplines.
- Ask questions to encourage the formation of associations; for example, "What does this look like? Remind you of? Sound like? How is it similar to ... ? Different from . . . ?"
- Demonstrate how the new skill grows out of prior learning and will lead to future learning.

Principle 12. Associate Purposeful Movements with New Learning

Guideline: *Include purposeful movement in the teaching of new concepts to forge a body-brain connection.*

Rationale

It is magical to me how physical movements are stored in the cerebellum and become automatic. By automatic, I mean that they are not conscious or deliberate—the body just knows how to move. Think of riding a bike. Once we have learned how, we never forget. In climbing stairs, we know just how to slant our body so we don't fall backwards, just how far to raise our foot for the next step—which is why we trip when risers are not built to standard height.

If you are a musician, think of how you memorize a piece of music. At first, you read the notes carefully and may practice the difficult passages in isolation. After much practice, the body begins to remember. At this point, playing a song is a kinesthetic-cerebellar function, not a cerebral function.

This incredible ability of the body to remember should be utilized by embedding purposeful movement in learning as often as possible. By purposeful movement, I am referring to movement that directly relates to the concept learned, not fidgety or random movements like bouncing on a big ball or hopping during learning. Forming a number shape with the body is one example of a purposeful movement, and several more are suggested in the "Applications" section. What the body does during learning can be permanently embedded in the cerebellum, serving as a powerful avenue for learning and recall.

Applications

Here are some suggestions for incorporating purposeful movement into good teaching practice.

- Use hand motions, such as "my two hands" (see page 61), as often as possible during learning. Design hand motions that resemble the printed shape of an abstract concept (such as numerals or math operators) as often as possible.
- When learning a pattern, pair it with a patterned movement. For example, when counting by twos, the class can march around, leaning heavily to the right on each even number.
- Use movements for recalling the meaning of + and -. When you say "plus," bring your crossed arms to your chest, demonstrating that more are coming to you, mimicking the plus sign with your crossed arms. Similarly, when you say "minus," slash horizontally away from your body with your right arm, mimicking the horizontal shape of the sign as well as showing that the number is leaving you.
- Use full-body skywriting rather than writing in the air. Have students stretch their whole body into the shape of the number being written (see illustration, page 32). For best results with this type of skywriting, form a large number on the floor with masking tape. Have a child stand and try to duplicate the shape with the body, while you provide support so that he or she does not fall. Then have the child quickly write on paper what he or she felt in the body. The action of moving the body to form the number mirrors on a larger scale the movement the children will do when writing the number. Being able to see the number while body-tracing it integrates the visual and kinesthetic modalities.
- Have pairs of students make full-body numbers.
- Have students walk along a large masking-tape number on the floor.
- Engage the children in thinking of motions that remind them of the concept being learned.

- As often as possible, learn by the "see, say, and do" method. Give the children a visual image, let them say the concept, and get them to move in a meaningful way all at the same time.

CASE STUDY

During my early years of working with children, I did much of what I did because I thought it was "normal practice." I cannot say that I had a good reason for what I did. Take, for example, the daily routine of counting to 20 while one child pointed to each number in turn. It sounds mind-numbing now, but back then it seemed the thing to do, and the children didn't seem to mind the routine.

One morning, Peter's mother asked me if I'd taught the children (ages four and five) to count by fives. When she had tucked Peter into bed the night before, he suddenly began counting by fives. After a great deal of thought, I finally understood how he had learned this. When it was Peter's turn to point to the numbers, he heard the numbers, said them, and pointed to them—but beyond that, each time his finger reached a new "five" number, his body would turn to the left in order to start another row of numbers. "Through the back door," Peter learned to count by fives. I think his body taught him this skill by emphasizing the numbers he said each time he turned to the left.

Principle 13. Teach from a Global Perspective

Guideline: *Teach new material from within a global whole and always specify goals for learning at the beginning.*

Rationale

I have touched on the global nature of this right-brained approach to teaching. Eric Jensen, in his book *Teaching with the Brain in Mind*, recommends starting with the global view before working through the sequential steps. An analogy would be telling your class, "We are going to Chicago. Now here are the step-by-step directions." Not only is a global approach essential for those learners who simply must know where they are headed, but it is essential for pattern discovery. (See the cartoon on page 12 to see how children respond to step-by-step directions without a final goal stated first.) One cannot discover patterns without seeing the whole. Another benefit of the global approach is that it lets the students know exactly how much they need to learn. The list of facts does not seem to stretch out endlessly, in an interminable string that can never be learned. Instead, the students know, "This is all, and there is no more." The global learner needs this information.

Applications

Here are some suggestions for teaching from a global stance:

• Show the goal (such as learning numbers to ten) before starting the lesson.

• Show the children the whole of what they will be learning such as Stony Brook Village (see page 7) or at least the whole street of problems they will be learning.

• Identify how each problem fits within the global whole, such as "On Tenth Street, 0 + 10 is the first house on the left, while 5 + 5 is the last house on the right" (see page 7).

• Encourage students to explore how each problem relates to problems on either side, noting similarities and differences. Brainstorm possible reasons for these similarities and differences.

CASE STUDY

I recall working with my group on math one morning. I had carefully chosen individual children's work based on where they were in their learning process. As I distributed the work, I noticed that Ben's face froze. He had taken one look at his paper and had a little anxiety attack. Casually I drew him aside to discover the source of his fears. He was able to verbalize that there were so many problems, and he was sure he didn't know them all. I knew something he didn't know, however, namely that every single problem on the page was a variation of these three number families: 1 + 6 = 7, 2 + 5 = 7, and 3 + 4 = 7.

Instead of telling him this, I gave him seven counters and two bowls. I asked him to find out how many ways he could distribute all the counters into the two bowls without repeating any combination. He got right to work, assuring me confidently that there were "lots" of ways. He quickly found the three families, then began to falter, although still insisting that he would find more ways. I let him gnaw on that particular bone for a while. When he finally gave up, I asked him then to look at the problems on the practice sheet. Because of his brief discovery time, he quickly recognized that indeed he did know all the problems and that they were just variations of each other. Fully at ease, Ben confidently completed his practice sheet. What made the difference? The "all and no more" approach. All of these problems are derived from those three families:

1 + 6	7 - 3	2 + 5	7 - 1	3 + 4	7 - 4
6 + 1	7 - 6	4 + 3	7 - 2	7 - 5	5 + 2

Conclusion: Teach to the Children's Brains

In a nutshell, everything we know about how a child learns may be discovered by observing children as we work with them in the classroom day in and day out. If an approach is not working, of course we change it. If a child experiences distress or shuts down, we stop what we are doing. If one method is not working, we create another avenue to understanding. The bottom line is how determined we are that all of our students will learn, remember, and grow into mastery and confidence. This goal should matter more than anything. There is absolutely nothing like giving a child's brain what it likes! The traditional method of teaching children is rather like pushing a heavy lawnmower up a steep incline, whereas a method that is compatible with children's brains is like watching a robot-driven mower do the work while lazing on the porch sipping lemonade.

3 UTILIZE ASSESSMENTS to SUPPORT LEARNING

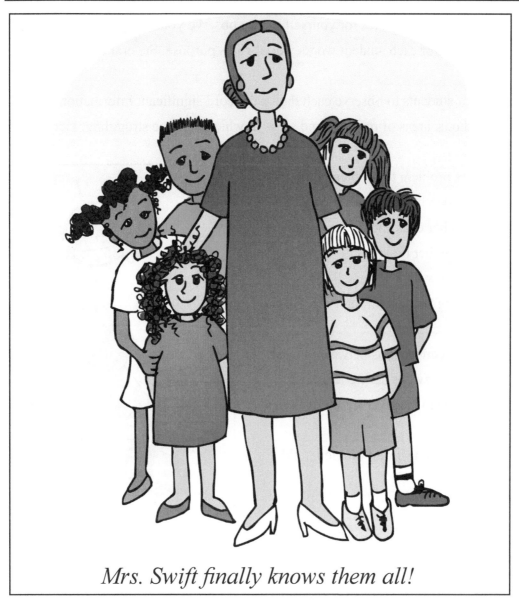

Mrs. Swift finally knows them all!

In order for the right-brained method to work successfully, a new view of assessments will be combined with the elements of good practice outlined in chapter 2. Underlying everything we do is a deep respect for children that leads us towards practices that support each learner. Respecting children means assuming the responsibility for taking them to the next level in their learning. This attitude is foundational to success in teaching all children.

Respecting our students and assuming responsibility for their learning causes a significant shift in our view and use of assessments. Rather than being utilized for grade collecting, labeling, and evaluating the child, assessments become tools that teach us about our students - what they understand and where they need more practice.

INFORMAL ASSESSMENTS

The heart, the passion, drives all else. If you have the passion to ensure all your students learn, become a kid watcher. Notice what engages the children and what shuts them down:

1. Develop the habit of making notes for yourself as you observe your students at work. A clipboard or a notebook with a section for each student works well for this purpose. Several tracking forms are provided in Appendix B.

2. Target a handful of students to observe each day and record significant information regarding their work habits, learning methods, areas of mastery, and areas in which they are struggling. Here are examples of notes you might make:

- Toni compares problem to problem as she works.
- Juan uses "my two hands" as a strategy.
- Lateisha closes her eyes and draws on an imprinted image in her brain.
- Paul is struggling with 3 + 4 and 3 + 5.
- Alicia is hurrying too fast through her work to read the problems correctly.
- Lynn is making connections between this set of problems and the previous set we learned.
- Takeshi struggles to discover a pattern even after many repetitions.
- Eric looked back at a similar problem he'd already done.

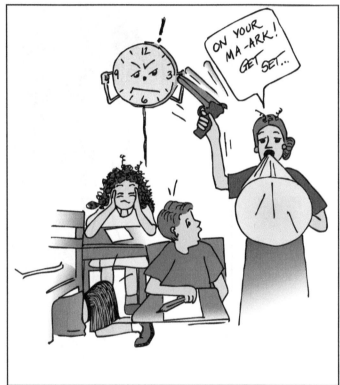

FORMAL ASSESSMENTS

Here are some suggestions for using written assessments.

1. Evaluate the Usefulness of Timed Tests

For the life of me, I cannot imagine a time in any child's future when being able to compute rapidly will save him from some catastrophe. What is the purpose of timing our students? Granted, there are some children who rise to the challenge of a timed test. Those children grow into adults who pay a lot of money to jump out of a tiny airplane wearing a parachute or leap off a bridge with a slender rubber band to prevent them from plunging to certain death. These people face challenges with an adrenaline surge and excitement.

For other children, timed tests represent a threat that interferes with their ability to think. The result of timing these children is that they freeze up, do poorly, and eventually experience tremendous anxiety around test taking. There are still other children who are wired in such a way that they simply cannot hurry. They are methodical, thorough workers who love to check their progress as they go along. They answer one question, adjust their body in the seat, look back at the well-answered problem, deliberately focus on the next problem, and repeat the pattern. Trying to make this type of child hurry will only result in frustration for the teacher, not in increased speed from the child. These are the students who may complete only the first half of a test, but the problems they complete are worked perfectly. Yet they receive a failing grade. Does this grade reflect their knowledge or their ability to compute? No. All it reflects is their speed, which would be much better determined in a footrace.

CASE STUDY

One day, I watched my children doing a sheet of double-digit subtraction problems. I watched particularly closely the progress of two boys who sat next to each other. One, barely five years old, quickly scoped out the nature of the work and literally looked and wrote his way down the page. He worked with energy and enjoyment. He "rabbited" his way through the page with amazing speed.

The other boy, age six, also had an air of being at ease. He worked the first problem, wrote the answer in his best handwriting, looked at his pencil point, looked back at the first problem, looked up and smiled at me, wiped imaginary eraser crumbs off his paper, then turned deliberately to the second problem. He "turtled" his way through the paper. He was not wasting time, nor did he daydream. He was being thorough and was proud of his work.

When I checked both papers, my rabbit missed five problems, which he easily corrected, while my turtle got a perfect score. Both knew how to do the computation. Both children are beautifully made. Both are smart. Both felt competent. But they worked very differently. Some people are not meant to be rushed.

2. Avoid Assessing the Whole Class Simultaneously on the Same Material

In order to administer fair formal assessments that accurately reflect each child's ability and knowledge, all the children would have to be at the exact same place in their learning processes. This is not a reasonable expectation. Remember Alice, who took months to master 1 + 2, then only four weeks (once I found her language) to master all the rest of the problems to ten? If I had administered formal assessments while Alice was in her period of discovery, she would have received failing scores. This failure would have communicated clearly to her that she could not do the work, taken her focus completely off her learning process, and diverted it to the process of test taking. Worse yet, her self-esteem would have suffered as she began to focus on comparing herself to the other students.

3. Design a System for Self-Pacing

Who is to say that a child must master computation to seven by the date set in the curriculum? Does it really matter when mastery is achieved? An important question to ask regarding such a schedule is, "Is this best practice for the child, or is this for adult convenience?" Again, if our goal is to educate, our measure of practice will be the child and his or her needs. Instead of testing all the children simultaneously, design storage cubbies in your room containing tests that the children can take as they are ready.

4. Give a Grade Based on Three Assessments

By averaging the results of three assessments, we can ensure that a particular score is an accurate reflection of knowledge, not a poor night's sleep or a bad day. Counting scores from three sessions also takes the pressure off the child and the focus off grades.

5. Use Assessments to Identify Areas Needing Practice

Use assessment results to identify children who need practice in the same areas. Group children with similar needs at math centers. Enlist parent or community volunteers to help lead practice sessions.

6. Record Progress and Mastery for Each Child

Progress charts geared to the method used in this book are provided in Appendix B. They are designed to yield information that is valuable both for you as a teacher and for sharing with parents. You will be able to tell at a glance where each child is in his or her learning.

7. Use the Progress Reports to Make Partners out of Parents

Focus on keeping parents informed of mastery, not grades. Work to create an atmosphere of being team members working together for the child's benefit.

Part II - The Method

Chapter 4: Learn Numbers

Chapter 5: Understand the Meaning of Numbers

Chapter 6: Make the Transition from Visual to Symbolic

Chapter 7: Master Computation to 10

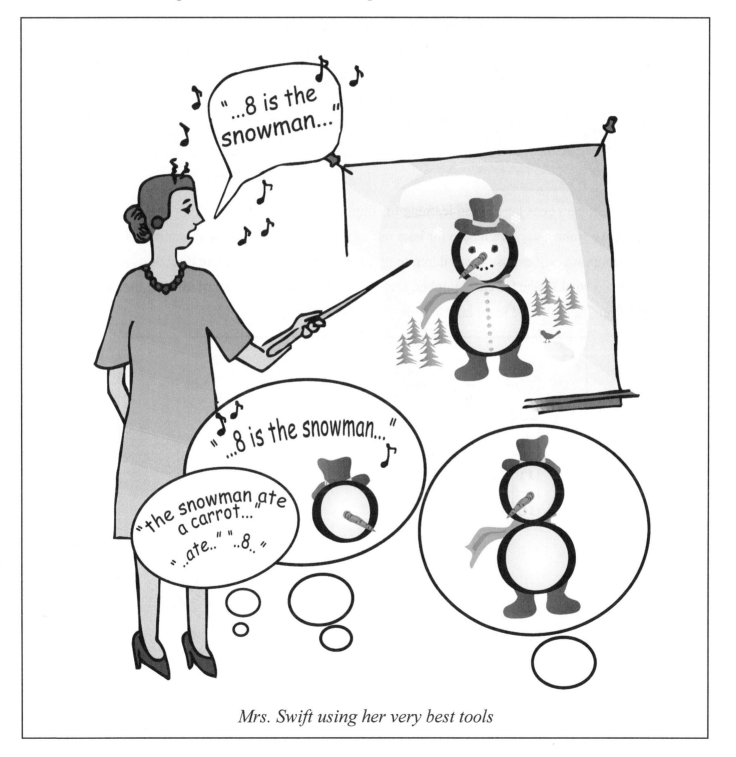

Mrs. Swift using her very best tools

4 LEARN NUMBERS

Goals for This Chapter

1. To recognize numbers to ten (and to recognize the numerals that make up the Arabic number system)
2. To count to 20
3. To order numbers sequentially from one to ten (or 20, for those who are ready)
4. To supply missing numbers from within a global pattern
5. To discover patterns within a global whole
6. To write the numbers from one to ten correctly

Learning numbers goes far beyond learning to count and recognize the numbers on sight. In this chapter, we want to lay a rich base for later use of numbers. Consistent with our right-brained approach to learning, we will expose the children to numbers from several different directions at once. Students will get a visual image for each number, see each number in context, learn to write the numbers correctly right from the start, and learn how to order numbers correctly.

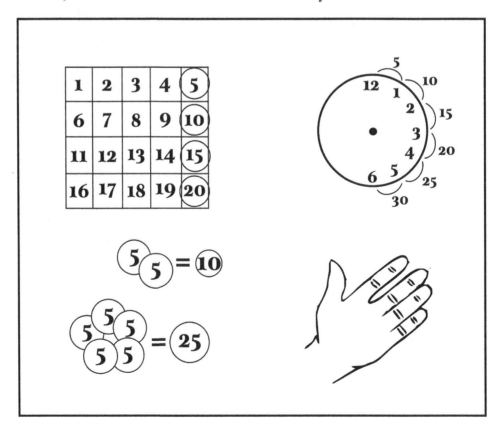

I arrange the numbers in rows of five from the very beginning. This arrangement is useful through basic computation and into multiplication and division. Consider all the applications of units of five: they relate directly to telling time in five-minute increments and to counting money. The set of five also corresponds to the number of fingers on the child's hand (see illustration page 40). So we will start in the same way we intend to continue, by arranging the numbers in a format that will be used repeatedly later.

Next, we identify two rows of five. Ten is just as significant a group as five is. Ten becomes the next important anchor number. (An anchor number is a point of reference against which the location of other numbers is defined. For example, 12 is two more than ten.) There are ten fingers on two hands, and our number system is base ten, so this number relates directly to place value. We learn to compute to ten and use the exact method learned to that point for computation beyond ten. All of this is a deliberate attempt to connect the stages of learning in a smooth flow, where initial learning is small, easy, and general, and all new knowledge is added to this all-important beginning in the form of specialization (an application of principle 11 in chapter 2). So, let's get started! Refer to the chart on page 41 to see what the numbers look like.

MAKING A NUMBER CHART

MATERIALS NEEDED:
- paste or glue stick
- sheet of tagboard at least 25" x 14"
- lamination equipment or clear plastic shelf paper (optional)
- place to display in the classroom

Stylized numbers are provided in Appendix A beginning on page 80. These may be used for a number book or for a large classroom number chart. To make the chart, follow the directions below.

1. Cut out the number images that begin on page 80.
2. Paste in rows of five on a sheet of tagboard, as shown in the illustration on page 41.
3. If desired, laminate the chart or cover with clear plastic shelf paper for protection.
4. Position the chart prominently on a classroom wall so the children will view it often, and visual imprinting will begin.

USING THE NUMBER CHART: PATTERN DISCOVERY

The children will make discoveries on their own if they have time to look at the chart and talk to each other about what they see, but some guidance from you will jump-start the process.

1. Group the children around the chart. Ask them what they notice about it. They might point out that the same number appears more than once (for example, there is a 4 in 4 and 14), or they might just talk about the shapes of the numbers.

2. When the children have finished making comments, point to the 2 in the top row and ask, "Do you see another two?" (Point out the 12 two rows below.) Then discuss other, similar patterns:

- The 1, 6, 1, 6 pattern in the left-hand column
- The 2, 7, 2, 7 pattern in the second column
- The 3, 8, 3, 8 pattern in the third column
- The 4, 9, 4, 9 pattern in the fourth column
- The 5, 0, 5, 0 pattern in the last column

3. Be creative in your exploration of these patterns to help the students make associations in multiple

modalities. As often as possible, draw out discoveries by asking leading questions rather than just by telling the children what to look for.

4. Talk about similarities in the number shapes. For example, you might pose these questions:

"Do you think that maybe the 1 in the first row fell straight down and became curved at the bottom, forming the number 6?"

"Do you see a connection between the 2 and the 7?" (Both numbers contain a slant from top right to bottom left.)

"Look at the next column. How are the 3 and 8 alike?" (The 3 is just like an 8 with one side cut off.)

"Look at the 4, 9 pattern. How do the numbers look similar?" (Both numbers are top-heavy, with a line extending down to complete the number.)

"And look at the last column. What's similar about these numbers?" (Both numbers have a rounded part.)

If the number chart is extended, another pattern that will emerge is the 1, 1, 2, 2, 3, 3, 4, 4, etc. pattern in the tens place in each column.

LEARN THE NUMBER SONG

The music, lyrics, and suggested motions for the Number Song are printed on page 45. You may copy this page as needed. I have provided a very simple tune with guitar chords. If you have no musical training, you might ask the school music teacher or a musical friend of yours to make you a cassette tape of the music.

The song is obviously a musical/rhythmic approach to learning numbers, and the lyrics provide a linguistic association to help in recognizing number shapes. The song, once learned, will assist your students in reaching four of the six goals for this chapter. It becomes a self-corrective strategy children can use as they learn to count in sequence (goal 2), find missing numbers (goal 4), put numbers in the correct order (goal 3), and complete simple number-recognition activities (goal 1).

The movements accompanying the song are one way to add a kinesthetic element to learning, but the world is full of other fun ways for children to practice forming numbers. They can—

- write each number in the air while singing the corresponding lyric
- write each number on paper as they sing about it
- finger-paint numbers on glossy paper
- trace numbers cut out of sandpaper
- form numbers out of modeling clay
- draw numbers in a sand tray
- form numbers from pipe cleaners
- turn each number into an object they associate with it (as I did in the stylized numbers chart)

For a tactile experience, half-fill a large self-sealing plastic bag with tempera paint. If the bag is placed on a solid surface, such as a table or the floor, children can write each number on the bag with a finger. The number will appear momentarily as the finger displaces the paint.

OTHER KINESTHETIC ACTIVITIES

The following ideas address number writing, recognition, and counting goals:

• Use wide masking tape to create very large numbers on the floor. Let the children walk on the tape, starting at the top just as they would if they were writing each number.

• Children can use their bodies to form numbers by lying down on top of the masking-tape numbers. They probably will have to pair up to make a whole number!

NUMBER RECOGNITION GAMES

BINGO: Give each child a copy of a Bingo Sheet (see 5.3a & b on pages 110-111) and ten poker chips. As you call out numbers in a random order, have the children place a chip over each number you call.

(Continued on page 47)

The NUMBER SONG

45

Kinesthetic Strategies for Preventing Backwards Numbers

The numbers 2, 3, 5, and 7 are "cave numbers." These numbers have to be written so that Big Bear—which is the index finger on their left hand—can crawl in when it is time to hibernate. Thus, they have to draw the shape of the number around their left index finger as it lies on their paper. This strategy works only for students who are right-handed. For lefties, explain that their pencil point is Big Bear. When they are holding their pencil ready to write the number, they have to make the cave opening face toward the pencil point.

These numbers - 2, 3, 5, and 7 - are the ones most often reversed. This story provides your students with a strategy for correctly forming these numbers, and a tool for immediate self-correction.

Strategies to prevent number reversals for right-handers.

Strategies to prevent number reversals for left-handers.

For 4, 6, 8, and 9, demonstrate how they fit within the hand resting on the paper (see illustration). All four numbers begin with a movement toward the hand, followed by one away from it. Left-handed children can extend the thumb of the right hand as it holds the paper, then draw around the thumb to make these numbers (see illustration).

GO FISH: Divide the children into groups of about four. Give each group a deck of number cards (or a standard deck of playing cards with the face cards removed will work). Deal all the cards and explain that the goal is to make pairs of numbers. Each player in turn has to request a specific number from one of the other players in order to make matches. The player with the most matches wins.

GO GRAB: Make two sets of the numbers from one to ten written on tagboard. Select 20 children and safety-pin a number to each child's shirt. Have the children line up in two teams, in order from one to ten, and stand facing each other about four feet apart. Toss a beanbag on the floor between the two groups while calling one number. The two children, one from each team, who are wearing that number run to the center and try to retrieve the beanbag. (With 30 children, make an extra set of numbers and have the children stand in a triangle or circle.)

LINE UP: Pin a number on each child's shirt as described in the previous activity. Divide the children into teams of ten composed of one child wearing each number. Give the teams a specific amount of time to arrange themselves into the correct sequential order. (The children who are not wearing numbers can act as timers and checkers for the teams.)

STAND UP: Pin a number to each child. Sing the Number Song (page 45) in a group. As each number is named in the lyrics, have all the children wearing that number stand up. (It is not necessary to have an equal number of children wearing each number.)

HOPSCOTCH: Outline a hopscotch grid on the floor with masking tape. (Make sure each square is large enough that children can actually step in it.) To play the game, a child tosses beanbags into squares with certain numbers in them, then hops through the squares in order from one to ten, skipping over any box with a beanbag in it. Have the class say each number the child hops on.

BEANBAG TOSS: Use the hopscotch grid described in the preceding activity. Designate one child as the caller who will shout out numbers. As each number is called, the players try to toss their beanbags into that square. (Play with a small group or set up a rotation system so everyone has a chance to play.)

SWIFT SAYS: Use the hopscotch grid small group. Give one child at a time instructions. Here are examples:
- Hop once on three.
- Sit on nine.
- Put your hand on five.
- Jump twice on six.
- Yawn on eight.
- Put both feet on four.
- Stomp on seven.

ASSESSMENTS

This section suggests informal ways to assess achievement of the goals for this chapter. See Appendix B for reproducible tracking charts and progress reports. There are six reproducible worksheets in Appendix A that correspond to the assessments described in this section. When each child has completed all six pages, you could compile the work into a small booklet for the child to take home. (You may want to keep a copy of this booklet in each child's portfolio.)

Recognizing Numbers

Give each child a fives chart to ten with numbers supplied (4.3, page 100). Call out directions to mark or color individual numbers with specific colors. Give directions slowly enough that the children have enough time to follow you but not enough time to become carried away with coloring. Sample directions follow, or you can make up your own, as long as each number is marked in a unique way, each direction involves only one number, and you do not call out the numbers in sequence.

- Put a circle around the three.
- Color the nine yellow.
- Put an X on the five.
- Draw a line under the seven.
- Make a box around the eight.
- Color the four red.
- Color the two blue.
- Color the one green.
- Color the six orange.
- Color the ten pink.

Counting To Ten

Listen to each child count from one to ten. If the counting is correct, check his or her name off your assessment list (Appendix B, page 173).

Ordering Numbers Sequentially

Give each child a page containing the numbers from one to ten in random order and a blank grid. Have the children put the numbers in correct order (4.4a, page 101). Use worksheet 4.4b (page 102) for those children who are ready to order numbers to 20. They can either cut apart the numbers on the worksheet and glue them in the correct order on the grid, or simply write each number in its appropriate square.

Supplying Missing Numbers

Give each child a copy of 4.5a and 4.5b (pages 103 and 104). Have the child fill in the missing numbers to complete the number grid. There are five different choices, three for numbers to ten and two for numbers to 20, so you will have options for practice or repeat assessment.

Discovering Patterns

Give each child a five-frame chart to 20 (4.6a, the top half of page 105). Have the children color as many patterns as they can find (for example: 1, 6, 11, 16).

Writing Numbers

Have children write the numbers to ten in their best handwriting. Allow them to choose a favorite color of crayon with which to write. (Use worksheet 4.7, page 106, or for children who can write the numbers to 20, use worksheet 4.6b, the bottom half of page 105.)

EXTENSION PROJECTS

You may wish to have children create personalized number books (see pages 124 - 125). These books could be as simple as pages of the stylized numbers, or there could be several pages for each number. For example, for the number three, the book might contain a page of the stylized numbers, a page of threes written by the child, a "found objects" page (such as a leaf with three lobes), a page containing a drawing of the child's three-member family, and a page of three fingerprints which the child has transformed into critters by drawing with a felt-tip pen.

The class could cooperate in making a set of ten books, one for each number. Each child could contribute one page and choose his or her favorite way of representing the particular number. Here are examples (refer to Gardner 1993 for explanations of the intelligences):

- A verbal/linguistic child could dictate a story about three of something.
- A logical/mathematical type may want to construct objects made from three shapes.
- A child strong in the naturalist intelligence may choose to find or draw objects from nature that have qualities of the target number (an ant with three body parts, for example).
- An interpersonal child may want to draw the members of his or her family (if they represent the target number) or draw the target number of close friends.
- A musical/rhythmic child could create a short poem, rhyme, or song about the number.
- A kinesthetic child might want to create a dance to go along with the song or rhyme the musical/rhythmic child wrote.
- An intrapersonal child might want to dictate a statement about what each number reminds him or her of, as a verbal "portrait."
- A visual/spatial child might be interested in drawing or painting original stylized numbers (different from those included in this chapter).

The ideas are endless, and the children themselves will come up with some of the best ones! It is wise to allow the children the choice of working alone or with a partner. The activities will result in many beautiful creations that will be admired and enjoyed for weeks. Parents certainly would enjoy seeing them displayed during an open house or parent/teacher conferences.

5 UNDERSTAND the MEANING of NUMBERS

> **Goals for This Chapter**
>
> 1. To visually imprint number meanings
> 2. To make visual connections for numbers to ten
> 3. To visually imprint computations to ten
> 4. To form visual imprints of where numbers are in relationship to one another
> 5. To recognize numbers within their context
> 6. To discover relationships between numbers

The magic ingredient in visual computing is the brain's ability to recall visual images. If you gain nothing else from chapters 1 to 3, the concept of visual imprinting is a must. In order for this method of computing to work, visual imprinting must take place—and it will, if you allow time to do some things that might appear at first glance to be just playing. The more care you take at the stage of the process described in this chapter, the more success your students will reap in future computation.

This chapter contains various activities to promote visual imprinting, visual connections, and pattern discovery. The results of this process will not be measurable using the objective assessments we have become accustomed to using. I will, however, detail how to tell when your goals for this stage have been reached. The process itself is so much fun that it may be tempting to rush through the activities, but please do not give in to this temptation! If you allow sufficient time to complete the visual imprinting exercises in this chapter, you and your students will see the payoff once you arrive at actual computation (described in chapter 7).

VISUAL IMPRINTING FOR NUMBERS TO TEN

The objective for this section is for every child to emerge with a clear visual image for each number. You will not, of course, be able to dictate what that image will be—all that matters is that each child emerges with an image that is unique to his or her memory. The visual image will probably involve a dot pattern and may include some color. You will have a chance to see some of these images as the children produce them for you in their final projects.

Initiating The Number Imprinting Process

Materials Needed
- overhead projector
- poker chips or similar tokens

If working with only a couple of children, use poker chips or math counters on a desk top.

1. Choose a target number. Set up the overhead projector and place that number of poker chips on it so the whole class can see them. *

2. Ask the children "How many are there?" Do not allow them to count; instead, ask them to guess! (We want this to be an instinctive, visual imprinting process, not a left-brained one.)

3. Quickly rearrange the chips and ask again, "How many are there now?" Repeat this step several times. You want the children to understand that no matter what the arrangement, the total number stays the same.

4. Ask questions to demonstrate that changing the location or arrangement of the chips does not change the quantity. For example, "If I put these on the floor, how many are there then?" or "If they are on the ceiling, how many are there?" "If I hold them tightly in my hand, how many are there?" Students will remember the point.

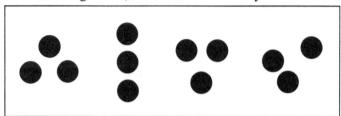

5. Place one dot grouping for the target number on the overhead projector. Ask, "What does this look like to you?" (Visual connections are being made.) Let volunteers give answers. Then show another grouping and ask the question again. Feel free to tell students what that dot pattern reminds you of. Here are some ideas to get your creative juices flowing:

bear's face	box	footprint	house	baby's chair	hill
building	mother's chair	truck	bushes	papa's chair	mountain

51

Imprinting for Odd Numbers

When you are teaching odd numbers, show a poker-chip arrangement in which all the chips but one are paired and say, "This reminds me of children walking with partners to the library." Ask, "Does every child have a partner?" When the students respond no, give each student or small group of students the target number of chips. Let students attempt to arrange the chips in a different way so that each "child" has a partner. If they decide that they are unable to make every child have a partner, suggest that the extra child must feel "odd." Discuss what the children might do to keep that child from feeling left out. We hope one of them will suggest taking turns! (This is a real-world connection that enhances memory of the concept.)

Materials Needed
- tempera or watercolor paints
- 1 sheet of sturdy paper per child, with a grid pattern on it (For fastest preparation, photocopy 5.1, page 107, one for each child.)
- fingers
- old clothes, apron, or painting shirts
- paper towels
- scissors
- lamination equipment or clear plastic shelf paper

Making Dot Cards

1. Using paper marked with a grid pattern, model how to make the target number of dots in each square by dipping that number of fingertips in the paint, then pressing those fingers in the square. Use a different arrangement each time. (You need not always use the same fingers, just the same number of fingers.)
2. Instruct students that each square in the grid should have a different grouping for that target number (visual imprinting). Stress creativity of design rather than speed!
3. Let the children talk about what they did and what they were thinking when they made the various patterns (visual connections).
4. When all the papers are dry, laminate them and let the children cut them into cards for use in the activities that follow.

Using Dot Cards For Visual Imprinting

You may wish to place sets of dot cards in the math center for groups to use during math time. See 5.2a-b (pages 108-9) for a basic set of dot cards or, preferably, let students share the dot cards they have made with the class.

Dot flash cards: Divide the students into pairs. One child displays a card and the other says what number it represents. Children take turns showing and naming the cards.

Bingo: Photocopy the dot card bingo boards. (There are two different boards, labeled worksheets 5.3a and 5.3b, pages 110-11.) Give each child a board and a set of chips. One child calls out the numbers while the others place chips on their boards.

Memory: Using two sets of dot cards, shuffle the cards and arrange them face down. (Use the children's dot cards or 5.2a-b.) A pair or small group of students can play memory. Each player turns over two cards in an attempt to find the two with the same number. Have the player name each card. If the cards have different numbers, the child must turn over both cards and the next player takes a turn. If the cards match, the player keeps them and takes another turn.

Go fish: Using four sets of dot cards, students can play go fish in groups of up to four players. Players must name the card they wish (rather than pointing or showing a card). If children use their own dot cards, then they will just be matching the cards with different arrangements of the same number of dots, rather than identical cards.

Floor dominoes: Make two copies of each page, laminate, and cut apart the floor domino cards (5.4a-f, pages 112-17). Let the children play in pairs or small groups, naming the numbers as they play.

War: Children play in pairs. Each player shuffles his or her dot cards and holds or places them face down. On a signal (such as, "1-2-3-War!") both players put a card down in the center at the same time. Whichever child has the larger number takes both cards. Alternatively, players can name a number before turning their cards over. If that number shows up, the player who said that number takes both cards. If not, the center pile builds. The game may also be played with naming "odd" or "even."

VISUAL IMPRINTING FOR COMPUTATION

The objective of this section is for the children to develop such a rich visual background for computation that they will "see" answers in their heads when confronted with abstract problems on a page. At this point, however, the process is quite passive. You will just be giving the students opportunities to look and discover, without actively learning math facts. The process of building visual images in children's memories is analogous to taking pictures of various groupings of dots. Until you develop the film, you cannot see the images, you just have to trust they are registered on the film.

Imprinting Computations

1. Copy the two pages of dot cards on 5.2a-b onto transparency film and cut apart.
2. Place one card on the projector and ask "How many are there?"

Materials Needed
- transparency film
- overhead projector
- activities 5.2a-b (pages 108-9)
- scissors
- marker and paper for each child

Repeat this procedure with each of the cards.

3. Give each child a marker and paper. Explain that the students will be reproducing on their papers what they see on the overhead. Use the following process for each number:

• Show the card.

• Have students close their eyes and "see" the card in their minds.

• Remove the card.

• Have students open their eyes and draw what they saw.

• Show the card again and have everyone name the number in unison.

4. Show a card on the overhead projector. Model a description of how the dots are grouped on the card and have the students repeat in unison. For example, "Two and one more is three" (see illustration page 29).

Using Dot Cards For Computation Imprinting

At this level, it is best to use identical pairs of cards, so I recommend making two copies of 5.2a-b, laminating them, and cutting the pages apart into cards.

Memory: This time when the children play memory, they will identify what they have paired by identifying the grouping, such as "one and two more is three." The objective here is to have the child move beyond counting to recognition of quantities as a prelude to addition. For the larger numbers, there is often more than one reasonable visual grouping. For example, the set of five dots in the bottom left card of resource 5.2a could be four dots in a square plus one more, or three dots in the top row plus two in the bottom. Any grouping that totals the target number of dots is acceptable.

Go fish: Players ask for a dot grouping they want, such as, "Do you have a one-and-two-more-is-three?" Any card with three dots qualifies.

VISUAL IMPRINTING FOR NUMBER RELATIONSHIPS

The activities in this section will lay the foundation for children's understanding of where numbers fall in relationship to each other within the global sequence of numbers. Just where is three in relationship to five? How much bigger is three than one? And how many more do you need to get from three to five?

Materials Needed
- pyramid dot cards (5.5a-b, pages 118-19)
- five-frame dot cards (5.6a-b, pages 120-21)
- photocopier
- lamination equipment or clear plastic shelf paper
- scissors

Make one photocopy of both resources for every two children. Laminate them and cut apart.

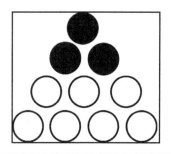

Pyramid Dot Cards (pages 118-19)

The patterns in the pyramid dot cards show how each number relates to ten. In this example, we can see a positive (black) three, a negative (outlined) seven, and ten as a whole. We also see three being made from one and two, and seven from three and four.

Five-Frame Dot Cards (pages 120-21)

This pattern anchors numbers to five and ten. This is seven. It is also five and two more, or "a hand and two fingers." It is also three less than ten.

Using the Pyramid Cards

Memory: Play in the traditional way except that children must describe each card as they turn it over, in terms of how many black dots, how many outlined dots, and how many total dots are on the card. For example, a pyramid with three black dots and seven outlined dots could be described as "three and seven more makes ten" or "seven and three more makes ten."

War: The children must label each card they lay down in terms of how many black dots, how many outlined dots, and how many total dots are on the card. For example, a pyramid with three black and seven outlined dots could be described as "three and seven more makes ten" or "seven and three more makes ten," depending on whether children want to play that the number of black dots or the number of outlined dots determines who wins.

How many: The first child turns over a card and asks the second, "How many (black) dots are there?" The second child answers and asks, "How many to ten?" The first child answers (by identifying the number of white dots). Then the second child turns over a card and the game continues.

Using the Five-Frame Cards

1. Begin with number-recognition activities. Show a card and ask, "How many are there?" Do this activity as a class first, then let the children pair off and play together.
2. Next, make a connection to children's hands. First ask how many fingers students have on one hand. Then ask how many boxes there are in the top row of the fives frame. Make a solid connection between the fingers of the left hand and the boxes in the top row of the chart. Next, make a solid connection between the right hand and the second row.

3. Hold up a card or place a transparency of the fives frame on an overhead projector. Ask the children how many dots there are, then ask them to show you on their hands how many.

This frame depicts seven. To show it on their hands, the children would hold up all five fingers of the left hand and two fingers of the right hand while saying in unison, "Seven."

4. Use the cards to identify the relationship between a number and its anchor numbers. (Anchor numbers are multiples of five, the numbers at the end of each row.) Do this as a class until the children are fluent with the activity, then let them pair up and practice. For example, given the preceding five frame (which shows seven), ask questions such as these:

- How many are there?
- How many more than five? (This is an example of using an anchor number.)
- How many more to ten?

To go one step further ask, "What if I put two more dots on? How many then?" You will hope to see that at least some of the children can figure this out without counting.

PATTERN DISCOVERY

The purpose of using a large five-frame chart is that it places the numbers in context, so that the relationships between them are easier to see and it is easier to discover patterns within the global whole. Using the five-frame chart (see illustration) will also provide a foundation for multiplication and division later on.

1. Make a copy of the five-frame chart to 70 for each child (5.7, page 122, which will supply charts for two children).

2. To begin, allow the children plenty of time to explore the number chart. Pattern discovery is enhanced if children are allowed to talk to each other about what they see.

3. Students could use colored pencils or highlighting markers to color in the patterns they discover. As discussed in chapter 4, each column forms a repeating pattern (see page 41). There are other discoveries to be made as well.

1	2	3	4	5
6	7	8	9	10
11	12	13	14	15
16	17	18	19	20
21	22	23	24	25
26	27	28	29	30
31	32	33	34	35
36	37	38	39	40
41	42	43	44	45
46	47	48	49	50
51	52	53	54	55
56	57	58	59	60
61	62	63	64	65
66	67	68	69	70

4. Let small groups or the entire class collaborate to make a really long chart (use the blank charts on 5.8, page 123), the scope of which they will determine. They may wish to find out if the patterns they discovered will keep on going or will stop at some point.

5. Lead a "what if game, using the five-frame chart and poker chips, to help students predict the patterns. For example, you might say, "Put a chip on five. Now count five more and put a chip on that number. What if you kept counting five more and putting a chip on each of those numbers? What do you think you would see?" Encourage student predictions. Now see what patterns develop when you add tens repeatedly.

6. Children who are very visual may continue finding patterns as they use this chart for discovery. For example, they might note that if they add five to any number, they get the number right under the number they started with. If they take five away from a number, they will end up with the number right above the number they started with. If they add ten, they will jump down exactly two rows. Also note that adding moves one down the chart, whereas subtracting moves one upward. Some highly visual children will apply patterns such as these in solving arithmetic problems, as exemplified in the case study of Ethan on page 28.

Becoming Familiar with Numbers to 20

To familiarize the children with the numbers to 20, use just four rows of the five-frame chart. (See 4.6, page 105, for a five-frame chart to 20 and a blank frame.)

1. Let students write the numbers to 20 into a blank frame.

2. Let them talk about the patterns they remember, if they wish to do so.

3. When the children seem familiar with the number frame, provide each with another blank frame and a set of poker chips. Play a bingo-like game in which you call out a number and the children place a chip on the blank space where that number would be found. Alternatively, as you call out each number, you could have students write it in the space where it belongs. Another variation is to identify a number by location rather than name. For example, "Put a chip on the number under three." "Put a chip on the number over nine." Your goal in these activities is to help the children become very familiar with each number's location on the "global map," its "address" if you will. When the fun of playing these games has begun to fade, it is time to move on into computation. I'm ready, aren't you?

FINAL PROJECTS

"When You Say ... I See ..." book: To make the booklet, cut sheets of paper in half, then fold five half-sheets in half. Place a cover page (see page 124) on top of the half-sheets and staple them all together on the fold to make a booklet with ten inside pages. Each student can write his or her name on the front cover, and decorate if desired. On each inside page, they write one number in order from one to ten. Have all the children turn to page 1, and ask, "When I say 'one,' what do you see?" Let each child draw his or her visual image of the number. Repeat for each number up to ten. (I would not recommend doing all the numbers at one sitting unless the students really want to.)

"Dot Critters" book: This book starts out looking like a "When You Say...1 See..." book. However, it uses dot patterns to represent each number. When the children have written the numbers and dot patterns, they take fine-tip markers and add legs, antennae, spots, eyes, smiles, or whatever to turn the images into critters. To add interdisciplinary learning, each child could dictate a story about the critters. For example, perhaps the child will add one more critter on each page because they are going to the playground and so gathering friends along the way. (A reproducible cover for this book is provided on page 125.)

ASSESSMENTS

Conduct informal assessment by observing the children daily during their partnered activities. Make notes to yourself about what you observe, including particular children's preferred method of remembering. Which children prefer the five frame and which the pyramid dot cards? Do they recognize dot patterns with ease, or do they have to think for a second before answering? Progress charts relevant to the skills in this chapter are provided in Appendix B.

The following procedure should not be used for giving grades, but rather serves as an assessment of students' learning thus far. You will be able to identify any weak spots in teaching or areas where the class needs practice. Make the assessment as low-key as possible; for example, don't describe it to the children as a test. The procedure is as follows:

- Choose a set of five-frame dot cards on overhead transparencies.
- Have the children number their paper according to how many cards you have.
- Place one card on the overhead at a time. Ask the children to look, close their eyes and visualize the numeral that corresponds to the number of dots on the card, then write the numeral on their paper.

You may repeat this assessment as needed to check for progress.

TRANSITION FROM VISUAL TO SYMBOLIC

> **Goals for This Chapter**
>
> 1. To make the transition from visual to symbolic representation
> 2. To understand the action of computation
> 3. To provide a brain/body tool for computation

This chapter continues the visual-kinesthetic approach we have begun but makes a natural transition from visual to symbolic representation. A great way to do this is through a storytelling approach. Stories allow the child to engage in the process, show the real-world applications of math, and begin to create an understanding of what is happening during computation. By this I do not mean memorizing a formula, but rather seeing and doing the action of computation.

STORY PROBLEMS

Twitch:

My favorite series of stories involves a precocious squirrel named Twitch. I have spent some lovely hours relating to my class the misadventures of this character, who is nine parts rash to one part cautious. Twitch's goal in life is to collect acorns, which he gathers in a little bag slung over his shoulder. Twitch lives in a hole up a tree in a neighborhood much like your neighborhood. There is a huge calico cat named Irving living on one side, and a hound named Chunk on the other. These neighbors add some twists and complications to the hunt for acorns. There have been times when Irving has pounced upon Twitch and caused him, in his fright, to drop acorns. At other times, Twitch, believing Chunk to be sound asleep, creeps too close to grab an acorn. Chunk wakes up, and the chase is on. As I tell a story about Twitch, the children pick up the number of acorns that Twitch finds, and put back any he drops. I purposefully use very small numbers so as not to divert a lot of attention to the counting. (The children might have little cloth bags and real acorns that have been coated with shellac or polyurethane to protect against rot and bugs—or at least small pebbles to represent acorns.) The children enjoy recapping the story when I have finished. They dump their acorns out on the table in front of them and talk about how many Twitch might have had if he hadn't dropped any!

Pool party:

Children at a pool party have a variety of reasons for getting into and out of the pool. The children can draw pools on paper and then use miniature plastic bears or other markers to represent the children in the story. At first, they can just practice taking the right number of children out of the pool and putting others in. As they do so, they will be feeling in their bodies the movements of adding and subtracting as motions of "bringing in" and "taking away."

At the playground:

Another possible setting is a playground at school (or another setting where children like to be or go). The focal point is a long swing set with children coming and going for various reasons. Again, have the students draw a representation of the swing set and give them materials to represent the children in the story.

Using Story Problems in the Transition to Symbols

Once the students are really clear on the movements behind adding and subtracting, the next step is to have them write equations as you tell the story. Begin by explaining the symbols in an equation:

- The first number in the equation is the number you have at the beginning of the story.
- The symbol (+ or -) shows that something happened: either some went away (make a horizontal motion with your arm that resembles the shape of the minus sign) or some came to join the first number (cross your arms over your chest to resemble a plus sign).
- The next number tells us how many either came (+) or went away (-).
- The = sign is the signal that we are at the end of the story and need to stop and take a picture of how many are left now.
- And the last number tells us what was left.

At first, talk about and model how to write the problem as you tell each story. Eventually, the children will join you in writing the problems during the story. As an example, if three children are in the pool, and two more get in, the expression is $3 + 2 = 5$. As you write this, you would use language such as this, "Three children were in the pool (point to the three) and here come (cross your arms over your chest) two more! Stop for a picture (point to =). There are five all together at the end of the story."

If you tell them next that one child wanted a drink and got out of the pool, the students would write $5-1 = 4$. You would explain: "There were five children in the pool. Oops! There goes (slicing arm motion) one child! Stop and take a picture. There are four left in the pool at the end of the story."

Story Problem Activities

For a gross motor approach to story problems, the children can act out the equation instead of writing or moving small objects.

- If the story is about a pool party, use masking tape to mark off the boundaries of the pool and have children actually jump into or climb out of the "pool" while you tell the story.

• If the story is about the playground, arrange chairs in a line to represent a swing set, and have children actually come and go as you tell the story.

• Another idea is to place a tape recorder and an audio tape of short recorded stories in the math center. Then children can practice computation on their own or in pairs during their time in the math center.

MY TWO HANDS

Here is the brain/body strategy that started Alice humming with computation. The skill it builds is that of enabling the children to see the sums, rather than counting up to them or memorizing answers. The reason to avoid counting up in computation is simply that if children learn this way, they will never progress beyond counting on fingers or on touch points on the numbers—a practice that is very common in our schools. To compute visually is to compute quickly.

To introduce "my two hands," hold up two fingers and ask, "How many?" Spread the same fingers wide apart and ask again. Then, with your fingers spread, say slowly and deliberately, "One plus one is two" (see illustrations).

Two and no more.   One and one is two.

Repeat this process for three. Say, "Show me three fingers!" Students will hold up three fingers. Then say, "Show me three another way." Guide them, if necessary, into showing a visual of 1 + 2. Describe this as, "One and two more is three."

Three and no more.   One and two is three.

Four and no more.   One and three is four.   Two and two is four.

Four is the four fingers on one hand. Repeat the same process as before, using the same language. Have the children watch and imitate you as you move your middle finger from one side to the other to make 2 + 2 and 1 + 3 (see illustrations).

Five is an important number. Five is all the fingers on one hand, and five is also one row on the five-frame chart. Five is the number you add to for numbers over five, and the number you take from for numbers under five. It is a great anchor number. The sum 1 + 4 is represented with the thumb and four fingers. Two digits and three more make 2 + 3.

Five and no more.   One and four is five.   Two and three is five.

Six is the fingers of one hand and the thumb of the other. The sum 1 + 5 is described as a hand and one more. For 2 + 4, you have two thumbs and four fingers. For 3 + 3, you have three digits and three more digits.

Six and no more.   One and five is six.   Two and four is six.   Three and three is six.

Continue in this fashion for all the numbers to ten. Use your hands to show the various ways to make each target number. If the child begins to count up to determine the number of fingers you are holding up, you have a teaching moment on your hands. Ask him or her to guess what the number is.

Extension Activity

Once the children can "see and do" the sums to a particular target number, show them why particular combinations are the same. Take 1 + 2 and 2 + 1, for example. Hold up your three fingers to show 1 + 2 like this: ▶

Then, while the children watch carefully, flip your hand, without moving your fingers at all, so that now the fingers show 2 + 1 like this: ▶

Talk about how the fingers stay just the same, but they trade places. I like to do this part of the lesson holding my fingers horizontally since this visual will closely reflect the "families" introduced later (see chapter 7).

Tough Spots

Sums for seven and eight seem to require the most practice. It helps quite a bit to use the children's own hands to help them see 3 + 4, 2 + 5, 2 + 6, and 3 + 5. I would use the five-frame dot cards as additional visuals with the sums that cause difficulty. Remember to use each difficult sum as a teaching moment and an opportunity for additional practice. You can demonstrate to the class the relationship between the "my two hands" method and the five-frame chart by holding out your hands, palms nearly flat, one over the other in order to show a row and several more. They would look something like this:

Five plus three on My Two Hands and on a Five Frame Card

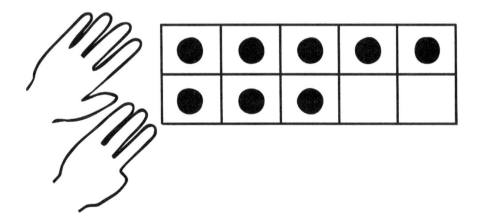

Making the Transition from Hands to Mind

Use your hands:

When I first introduced this concept to Alice, she quickly understood what was happening. I was delighted to watch her looking at her hands as she was doing her practice sheets. After a while, however, I feared that this practice could also become a crutch, although not as limiting as counting on one's fingers.

See it in your mind:

So I took Alice to the next level, which was to ask her if she could see her hands doing the movements in her head. She was game to try. She would squeeze her eyes shut and, after a moment, she would exclaim, "I can see it!" She must have, too, because she would quickly open her eyes and write the answer. It was not long before Alice rarely had to stop to picture a computation in her head.

7 MASTER COMPUTATION TO TEN

> **Goals for This Chapter**
>
> 1. To create a real-world context for computation
> 2. To provide a global view of all number families to ten
> 3. To conduct pattern discovery within number families
> 4. To master computation to ten

Now that students have laid a rich and vivid background for computation, let's get to it! Welcome to Stony Brook Village, the strategy I use to help students visualize computation! Let's take a walking tour before we start to work. A global view of the residential part of town can be found on page 7.

ORIENTATION TO STONY BROOK VILLAGE

Property Managers

You will be assigning students the duty of being property managers, so you will need to explain to them what this means. "Property," I tell them, "is what you can see in this picture: the houses and the land. Being a manager of all this property means that you take care of it." I explain that the job of the property manager is to keep track of who lives in the houses on each particular street. If a family moves out, the manager has to rent the house to someone else. The most important job of the property manager is to oversee how many people can live on each floor of the house. Even five-year-olds understand and use this title with ease. In addition, the title of manager reflects the children's own management of their learning process.

Streets and Houses

As you can see from the map on page 7, each street in the residential section is clearly marked with a sign. The name of the street matches the number written in the attic of each house on that street. The rule in this town is that only that many people can live in each house. So, only three people can live in each house on Third Street. On Fourth Street, only four people can live in each house, and so on. We don't know for sure how these rules were established, but we do know that over time, the rules got more and more involved and even a bit silly.

Apartments in the Houses

The next rule the Planning Commission made was that there were to be two separate apartments inside each house, so that each house could shelter two families, one on each floor. (The Planning Commission is a group of people who have meetings and sit around a table and make rules for the town. They like things done a certain way, and the property managers have to obey their rules!)

Apartment Residents

That rule was apparently not enough, because the next rule the Commission made was that no two houses could be populated in the same way. This means that no house could have the same pattern of residents as another house. For example, if you look at Fourth Street, you will notice that the first house on that street has no one living upstairs, but a family of four downstairs. The second house has to be different. It has one person living alone upstairs and three people living downstairs. In the last house, two live upstairs and two live downstairs. So you can't have two houses on Fourth Street with one person upstairs and three downstairs.

The First House on a Street

We are not certain if a rule was made to this effect, but the fact is that property managers always seem to leave the top floor of the first house empty if they have a family that has the maximum allowable number of residents. For example, if a family of four wanted to live on Fourth Street, the property manager would put them in the first house, on the bottom floor, so that the empty apartment could be at the top of the first house (see illustration).

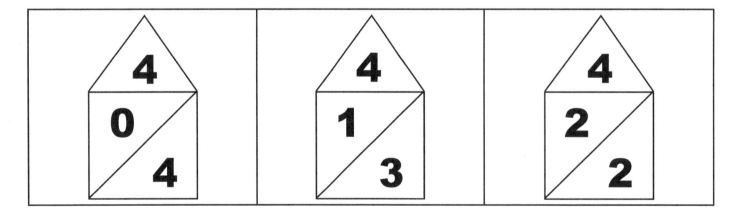

Moving

The funny part in all this rule making is that once the families have lived in their apartments for even a day, they are allowed to switch apartments with the other occupants of their house. They may not, however, switch houses altogether, as this would upset the delicate balance of the population. If one of the two families in the first house tried to trade places with a family in the second house, there would not be the right number of people living in either house anymore. The property manager would get into trouble with the Planning Commission, and everyone would be sad.

PATTERN DISCOVERY

Copy the village map (page 69) onto transparent film and project it on an overhead, and give a copy to each student. Guide students to notice interesting facts about the distribution of people in the homes.

Odd and Even Houses

• The even-numbered houses and streets are all on the right side of the map.

• The last house on each even-numbered street has an equal number of residents per floor. For example, the last home on Fourth Street has two residents living upstairs and two downstairs. If you look at the last houses on Sixth, Eighth, and Tenth Streets, their residents are evenly divided also.

• Odd-numbered streets are on the left side of the map.

• The house at the end of each odd street has one more resident on the bottom floor than on the top.

Comparing Streets

• Choose two streets, say Fourth and Fifth Streets, and do some comparisons. Ask, "How many houses are on each of these streets?" (Three houses on each street.) "Why do you think there are the same number of houses when five is bigger than four?" "What is different about how the numbers are placed in the houses on these two streets— particularly in the last house on each street?"

• Repeat this process with Sixth and Seventh Streets, noting that Sixth Street is an even street, so there are an equal number of people living on each floor of the last house (namely, three people on each floor).

• Seventh and Sixth Streets are remarkably similar. Take time to note that the top floors of the houses on those two streets are identical. The bottom floors of the houses on Seventh Street contain exactly one more person than the bottom floors of the houses on Sixth Street. Ask, "Why is this the case?" (The attic number on Seventh Street is one digit larger than the attic number on Sixth Street, and the extra person lives downstairs in each house.)

• Repeat these comparisons with Eighth and Ninth Streets.

Number Sequences

• Note the sequence of numbers in the last houses on Fourth, Sixth, Eighth, and Tenth Streets. The numbers of residents form this pattern: 2 + 2, 3 + 3, 4 + 4, and 5 + 5.

• Compare that pattern with the pattern of residents in the last houses on Third, Fifth, Seventh, and Ninth Streets: 1 + 2, 2 + 3, 3 + 4, 4 + 5.

Houses on Streets

• Note the number of houses on each street. For even-numbered streets, the number of houses equals half of the attic number plus one. Thus, on Fourth Street, there are half of four (that is, two) plus one more to total three houses. Fourth and Fifth Streets have the same number of houses because half of five is the same as half of four, except for one extra person, who lives on the bottom floor.

• See if the children find any additional patterns or points of interest, and record their findings.

PROCEDURE for TEACHING NUMBER HOUSES

Work through all the following steps for one street at a time before presenting the next street, starting with Third Street. Worksheets 7.4 to 7.10 refer to series of worksheets (distinguished with the letters a, b, c, and so on). In each cycle through these steps, present only the version of the worksheet that is appropriate for the number family you are targeting. Procedures are spelled out starting on page 70.

Discovery

1. Felt houses and bears: The children arrange small toys, such as plastic bears, in felt number houses, in order to discover with concrete objects the combinations that total the attic number (7.1 and 7.2a-b, pp. 125-129).

2. Smallest to largest: The children place number cards in the windows of the number houses, thereby discovering the pattern that, on each street, the smallest number goes with the largest one, the next smallest with the next largest, and so on (7.1-7.3).

Exploration

3. Discovery worksheet: Activity 7.4, pp. 130-133, mirrors the discovery process in step 2, but now the children write the appropriate numbers in the house windows.

4. Practice houses worksheet: Activity 7.5 has students repeatedly write the pairs of numbers for each street in order to gain automaticity.

5. All and no more: Students examine 7.5 to determine that there are no other possible combinations that total the target number.

6. My two hands: Children make the number combinations for the street on their hands in order to build a brain/body/visual connection to the number pairs for that street.

7. Practice strips: Activity 7.5 is presented again, this time with houses partially completed in a random order, rather than the strict numerical order of the Stony Brook street.

8. Families went walking: Activity 7.6, pp. 139-142, introduces subtraction from the target number.

9. Additional practice: If any students have not completely mastered the tasks presented so far, provide additional practice.

10. Relate houses to addition and subtraction problems: Activity 7.8, pp. 147-152, takes one street at a time, displays the possible problems for that street, and has the student refer to the houses to answer each problem.

11. Mixed addition and subtraction practice: Activity 7.7, pp. 143-146, contains mixed addition and subtraction problems for the target street.

12. Writing problems: Activity 7.9, pp. 153-158, is an optional but recommended worksheet that has children translate each house on the target street into the traditional addition and subtraction problems that correspond to it.

13. Mixed numbers practice: Once students have mastered at least Third and Fourth Streets, you can present worksheet 7.10, pp. 159-164, which mixes problems for the target street with problems for the previously mastered streets.

14. Additional practice: Activity 7.11, p. 165, is a page of blank number houses you can use to make up problems for any student who needs some extra practice before moving on. Activities 7.12a-e, pp. 166-170, provide for extra practice as well.

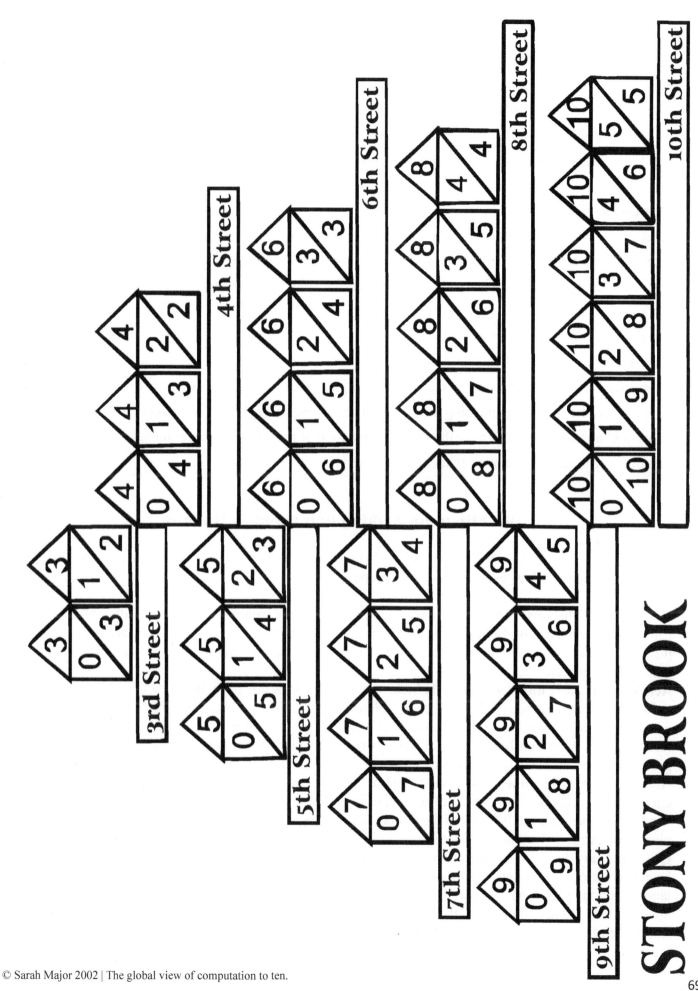

DISCOVERY

After discovering patterns for the whole village (the global approach), it is time to start working at the beginning and proceed in steps. The following steps are for each attic number from three to ten. Work through all the steps for one number at a time, starting with Third Street and ending with Tenth Street. (Note that there is no Second Street because in my experience, most children already know 1 + 1 and 2 + 0.) My rule of thumb is that students generally can work through steps 1-12 for a single street in one practice session.

1. Felt Houses and Bears

Students begin by using concrete objects to discover the combinations that total the target number. I like to use small plastic bears in felt number houses. Distribute to the children the correct number of houses they will need for the street you are working with. For example, for Third Street, each child will need the following:

- two houses
- two 3 number cards for the attic
- bears or tokens to fill both houses, six in all

Have the children place an attic number in each attic, so they remember how many bears can live in each house. Review the rules for populating the houses: The total number of bears (or tokens) must equal the attic number, and no two houses can have the same combination of residents in its apartments. Let the children figure out how to arrange the bears to meet these criteria.

Directions for Making Felt Number Houses

Activities 7.1 - 7.3b in Appendix A (pages 126-29) contain patterns for felt houses, attic numbers, and window numbers. Follow these directions for preparation:

Activity 7.1. Felt-house pattern: To make a template, place the pattern on a piece of stiff cardboard. Cut around the outside of the house and cut out the door and windows. Place the cardboard template on a piece of felt and cut around the perimeter of the house only. Use a marker to draw the windows, door, and attic line on the felt.

Activity 7.2a. Attic numbers—page 1: Make one photocopy of this page, laminate it, and cut apart the attic numbers. Place a piece of self-adhesive Velcro® hook-and-loop fastener on the back of each attic number so it can be stuck to the felt house. (Rolls of Velcro® are available in fabric stores.)

Activity 7.2b. Attic numbers—page 2: Make two photocopies of this page, then follow the directions given for page 1 to make laminated attic pieces to stick on the felt.

Activity 7.3. Window numbers: Make two photocopies of each of this page. Laminate the page and then cut apart the numbers. Place a piece of self-adhesive Velcro® hook-and-loop fastener on the back of each window number so it can be stuck to the felt house. There will be enough windows to make one town, plus a few extras. (These are used for step 2, following.)

Quick preparation: If the preparation time to make felt houses for all your students is prohibitive, you have two options. You might simply photocopy the house pattern, attic numbers, and window numbers; laminate all the pages and cut apart the attic and window numbers; then place Velcro dots on both the house pattern and the numbers, so the numbers can be stuck to the houses. Another option is to make only enough felt houses for a small group to use and to do this step with small groups in the math center, rather than with an entire class simultaneously.

2. Smallest to Largest

Next, put the bears away but have children keep the felt houses and attic numbers. Give them laminated window numbers. Have them line up the window cards in a row, starting with zero and finishing with the highest number they have. Then instruct them to figure out which pair of numbers goes together in each house, reminding them to use all the numbers. (Don't forget that for even-numbered streets, students will need two copies of the middle number in the series; that is, two 2s for Fourth Street, two 3s for Sixth Street, two 4s for Eighth Street, and two 5s for Tenth Street.) When all the students have finished, talk about what they have discovered. As shown in the illustration, the solution is to pair the first and last number in the row, then the

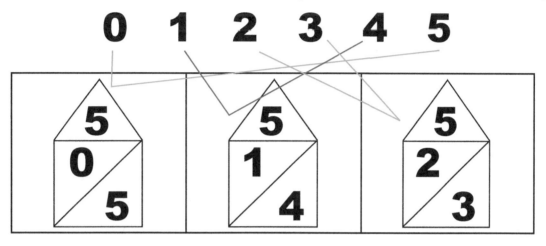

next two numbers in, and so on. By picking up and placing actual numbers, the children will feel the rhythm and movement of starting from the outside and moving in, particularly if they grab two numbers at once, one in each hand, to place in the windows.

Street Patterns

At this point, you will want to guide the children to arrange the houses in a meaningful pattern so they will be able to remember the whole picture without having to memorize the numbers. It would not be easy for children to remember the numbers in the houses shown above.

In contrast, they will have no difficulty at all remembering the numbers when they are arranged in sequential order like this:

Engage your students in a discussion about why this pattern is easier to remember than the previous one. See if they can discover on their own the sequence of numbers in the top floors: 0, 1, 2, which then continues from right to left in the bottom floor (3, 4, 5).

71

The Backwards C

A movement and shape that emerges in the number sequence is that of a backwards C (see illustration). This backwards C motion can provide a kinesthetic link to help the child recall the order in which the numbers are placed in each street of houses. Encourage the children to swing an arm in the backwards C motion as they look at their houses.

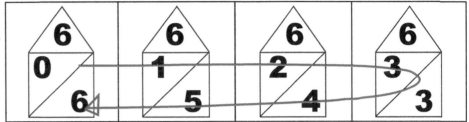

EXPLORATION

Now that students have had an opportunity to discover for themselves the patterns in the number houses on the street you are targeting, they will move to pencil-and-paper addition and subtraction tasks using the number houses.

3. Discovery Worksheet

Give each child a copy of the version of worksheet 7.4 for the particular number you are working on ("Figure out which families can live in each house," pages 130-33). This worksheet mirrors the discovery activity the children have just completed. Now, however, they write the appropriate numbers in the house windows rather than using preprinted numbers. Remind them of the self-assessment guidelines: They have to use every number provided, and they cannot use any number twice.

4. Practice Houses Worksheet

Locate and copy the version of worksheet 7.5 (pages 134-38) that has the correct number of houses you need for the street you are on. Make enough copies for the class and trim along the top and bottom dotted lines to make equal thirds. After the children write their names on their papers, have them turn the papers face down and fold the bottom flap up first on the dotted line. Then have children fold the top section under. When they turn the folded paper over, they should be looking at only the top row of houses.

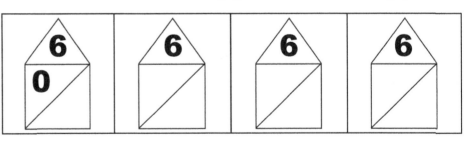

Tell them what the attic number is, and have them write that number in every attic. Then have them write a zero in the top window of the first house. (You could also write these numbers in on a master copy of the worksheet before copying for the class, as illustrated here.) Now have students fill in the numbers to complete the families on the street. When they have completed the top row, students should turn their paper over and complete the center section, then finish by doing the bottom section. Repeat this activity until the children can dash off the numbers without stopping to think about what to write.

5. All and No More

When the children have finished the practice houses worksheet, ask whether they have found all the possible combinations. When the class establishes that they have, point out that these few combinations are the only ways they will be able to divide that number. For example, for the target number six, they will make one house 0 + 6, another 1 + 5, then 2 + 4 and lastly 3 + 3, and those are the only four combinations they will be able to make that equal six. Remind them that a house containing 2 + 4 is the same as one that has 4 + 2.

6. My Two Hands

Do a quick review of the "my two hands" combinations for the target number (see page 65). At this point, encourage the children to make a sum on their hands (such as 2 + 4), then to close their eyes and "see" their hands showing 2 + 4. Thereby you are teaching them to draw upon imprinted pictures of the sums.

7. Practice Strips

Use the same practice house strips you used for step 4 (choose the appropriate version of worksheet 7.5, pages 134-38), but this time, before you copy enough for the class or group, fill in the upper window number on each house. Instead of following the established order for Stony Brook, write the numbers in a different order (see illustration). Explain that the families got tired of living in their apartments and decided to trade around on their street. The only rules in place now are that no two houses can be alike, and no more than the attic number can live in a house. I give students three sets of practice strips to complete, one at a time, so they cannot look at the problems they just completed to fill in the next ones. This activity will help the children move away from just filling in the "backwards C-shape" number sequence, and actually

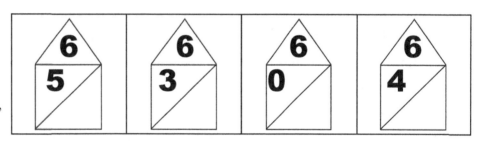

begin to pair sets of numbers together. A self-assessment can be "My Two Hands."

8. Families Went Walking

Locate the version of 7.6 (pages 139-42) that matches the target number you are practicing. This series of worksheets has an attic and upper window number, leaving the lower floor to be filled in (which is the operation of subtraction). Referring to the numbers provided at the bottom of the page, students choose which family is missing from each house and write that number in the empty window. The children may wish to invent reasons why the missing families left. Maybe they went to the store, the park, or Grandma's house. Your hope at this stage is that children can draw upon their memory of which numbers go together in each house without counting. If they need to look at their hands to figure out the answers, you may wish to stop and review steps 6 and 7 above. Alternately, let them refer to "My Two Hands" until they don't need to anymore.

9. Additional Practice

Depending on students' fluency and comfort level, either give more practice with the problems they are hesitant about solving or move to a full page of mixed addition and subtraction problems. For students who are confidently solving the problems without hesitation thus far, move to step 10. For students who need additional practice before moving on, make a copy of the relevent version of worksheet 7.5 (pages 134-38), write in appropriate problems, then photocopy enough for all the students who need practice.

10. From Houses to Number Problems

We will now begin the transition from the right-brained view of computation to the more left-brained, symbolic representation, using Activity 7.8 (pages 147-52). Having both views on one page is helpful as it will strengthen the connection between hemispheres in the brain. Explain to the children that all the problems on the page can be found in a house on the street. Each house can have two addition problems and two subtraction problems, which is why it looks like there are *so many facts to know!*

The purpose of this activity is to show the children that a single combination of three numbers can yield up to four different problems, but that if they know one trio of numbers, they can easily solve all four problems.

11. Mixed Addition and Subtraction Practice

Worksheet series 7.7 (pages 143-46) contains both addition and subtraction problems for each target number. (Present only the worksheet for the number family you are targeting.) Each mixed practice sheet targets only one sum, but because there are four addition and subtraction possibilities for that number, it is easy to provide a page of problems. Here is how I describe the problems to the children:

Subtraction: If the attic number is present and only one number is in the top part of the house, this means a family is missing (out for a walk, at the store, on a picnic, or whatever reason you come up with). The property manager has to figure out which family went walking.

Addition: If the two families are there but the attic number is missing, a big wind wiped out the attic signs, and you, the property manager, must replace the signs immediately. The attic number has to equal the number of people in the bottom family plus the number of people in the top family.

With practice, the children will simply become used to seeing the number families together. Some children remember them geographically on the street as they are problem solving (visual mapping), some resort to looking at their two hands, and some make other connections.

12. Writing Problems

Worksheets 7.9a-h (pages 153-58) take the transition to symbols one step further, as children change each house into traditional arithmetic problems. Present only the version of the worksheet for the targeted number family and work with the children as they need help in completing it. Discuss the various possibilities, reminding them that they can "lose" either attic numbers or families to make problems for each house. Discuss which symbol they would use for each type of situation. For the attic number six, and the house with two upstairs and four downstairs, they would be able to write two subtraction and two addition problems: 6 - 2 (by losing the bottom family), 6 - 4 (by losing the top family), 2 + 4 (by losing the attic number), and 4 + 2 (by losing the attic number). You might need to model completing this worksheet the first time you use it, but after that, the children should understand what they are doing for subsequent streets.

You have now completed the basic teaching procedure. If you have just worked through Third Street, go back to step 1 and start over with Fourth Street. Once you reach this point with Fourth Street, you can begin presenting practice sheets that have mixed problems for threes and fours, as described in step 13.

13. Mixed Numbers Practice

Once your class has mastered all the possible problems for Third and Fourth Streets individually, give them the mixed addition and subtraction practice sheets for threes and fours (worksheet 7.10a, page 159). Then, go back through steps 1-12 for Fifth Street, followed by Sixth Street. Once the children have mastered the Fifth and Sixth Street problems separately, you can present worksheet 7.10b (page 160), which has addition and subtraction problems for five and six, and worksheet 7.10c (page 161), which mixes problems for three, four, five, and six. Similarly worksheet 7.10d (page 162) has mixed problems for seven and eight; worksheet 7.10e (page 163) has mixed problems for nine and ten; and worksheet 7.10f (page 164) has mixed problems for seven, eight, nine, and ten. Present these worksheets as students master the targeted problems.

14. Additional Practice

Provide as little or as much practice with mixed problems as children need. You will find a page of empty houses that you can use to create individualized worksheets for extra practice (see worksheet 7.11, page 165). For example, you might present representative problems for all the numbers from three to ten in order to assess whether children have mastered them all.

You can target a variety of types of problems using worksheet 7.11. First make a photocopy of the worksheet and decide which number or numbers to target. Then decide which of the following type or types of problems you want students to practice:

- For discovery problems, fill in the attic numbers and let students write in both families.
- For subtraction problems, fill in an attic number and one family for each house, then let students write in the other family.
- For addition, fill in both families and let students fill in the attic numbers.

OPTIONAL TEACHING STRATEGIES

My rule of thumb is that students should be able to learn one street per practice session. If some of your students have difficulty learning certain streets, you will want to model making additional connections, so they learn to make connections on their own. Remember, the foundation of this system is that connections drive learning. Here are some connections between symbol and prior knowledge that have worked for students who had difficulty remembering the streets.

Label the Houses

I give associative names to the number combinations that recur on every street (see illustration on page 77). These four house types (Tornado House, Counting House, Parents' House, and Split House) account for every combination on Third through Seventh Streets.

Hand And ... Houses

Any house that has a five on a floor qualifies as a Hand And . . . House. Of course, a Hand And . . . House will at the same time be one of the four house types on page 77 (such as a Parent House—2 + 5—or a Split House—4 + 5). However, identifying the Hand And . . . Houses will help to tie those number combinations to the child's own hands (which are very familiar), as well as to the five-frame format that will be used for higher levels of math. The green houses in the illustration are the Hand And . . . Houses.

STONY BROOK

Notice that on each successive street, the Hand And . . . House moves one place to the right. Connect these houses purposefully with the child's hands, and use the "my two hands" language as you model the combinations on your hands:

- Five is a hand and no more.
- Six is a hand and one more.
- Seven is a hand and two more.
- Eight is a hand and three more.
- Nine is a hand and four more.
- Ten is two hands.

Make Up a Story for Each Street

In order to create solid connections for learning, have the children choose a family name for each street. For example, whichever street corresponds to the number of people in the child's own family might be named with that child's surname, and other streets could be named for other families well known to the child. Assume, for example, that Seventh Street is the Browns' street. Here is their story:

The Brown family has two parents, four kids, and a dog named Goliath that is very antsy.

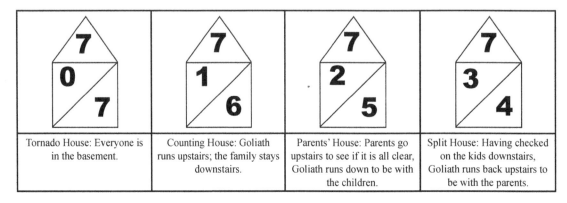

| Tornado House: Everyone is in the basement. | Counting House: Goliath runs upstairs; the family stays downstairs. | Parents' House: Parents go upstairs to see if it is all clear, Goliath runs down to be with the children. | Split House: Having checked on the kids downstairs, Goliath runs back upstairs to be with the parents. |

LABELED HOUSES

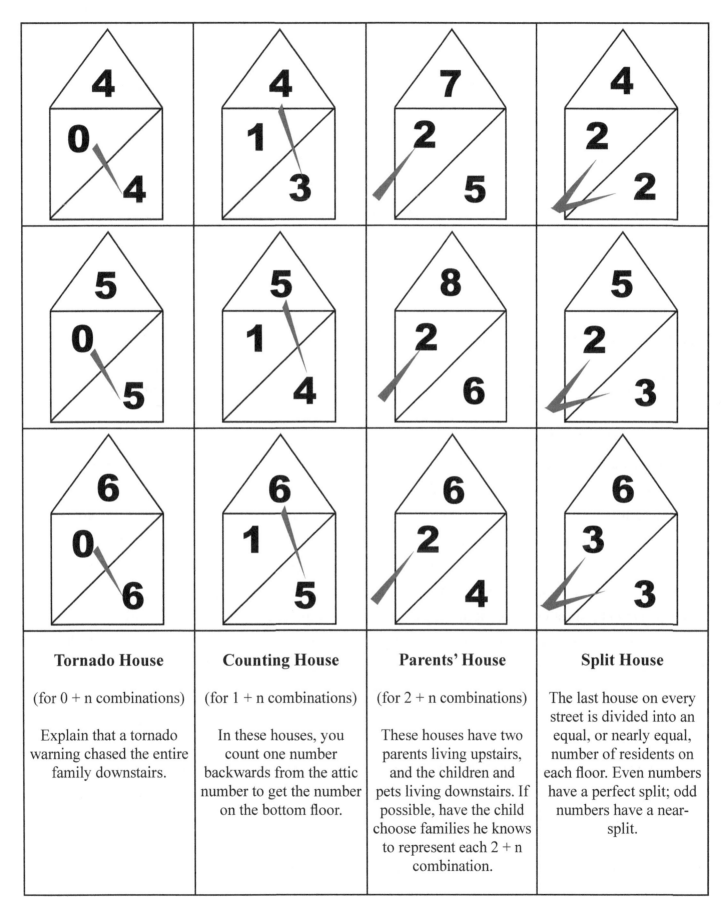

Have the child make up a story like this for every street with which he or she has difficulty. It is important that the child invents the family and the story, because the story is intended to provide a meaningful link to the number facts. A story you make up might have associations that are useful for you but not for the child.

Ask the Child

One of your most powerful tools for making connections will be to ask students repeatedly, "How can you remember this?" Doing this will prompt them to create mental images or mnemonic aids for learning and will give them the sense of being in control of their own learning. The mirror image to that practice is to ask repeatedly, "How did you remember that?" At first children might simply shrug and say, "I just did." Or "I don't know." But persist in asking. Before very long, they will begin to consider how they remembered, thereby rapidly developing powerful metacognitive habits that will strengthen their learning process. They will move from passive bystander to proactive learner.

Extension Activities

To strengthen understanding of math concepts, please find samples of extra practice worksheets in Appendix A (worksheets 7.12a-f, pages 166-70). Some are story problems, and others are blank so students can invent their own problems. At this stage, encourage the children to be as creative as possible in their work. It would be fun and a good learning experience to let partners come up with story problems they illustrate.

INFORMAL ASSESSMENT STRATEGIES

Observe the students as they work to determine whether any need further practice with a single street before moving on to mixed problems. Usually, children do not get confused. If they do, this only brings up a teaching moment. As you discover which problems cause difficulties, try strategies that relate the problem to its family or location on the street, rather than expecting the child to memorize an isolated fact.

The mixed numbers practice pages (worksheet series 7.10, pages 159-64) can also serve as a first assessment, if desired. Remember to use these worksheets to determine which problems need further teaching and which might need stronger connections. Remind the children, if necessary, that if the attic number is missing, they have to replace it because it tells everyone how many people can live in that house. If a floor is empty, that means one of the families went walking. Worksheet 7.11 (page 165), the blank practice houses worksheet, is also useful for making up a page of problems for assessment purposes.

CONCLUSION

Enjoy yourself. Encourage your students to decide now and then what math work they would like to do that day. I suspect you will find that they ask for specific streets on which they need practice. I would suggest keeping cubbies in the room with prepared sheets for each type of problem. On "free math day," students can simply choose whichever sheets they want to review. Allowing students to choose the math work they do is very empowering for them. They will delight in making decisions for themselves within the parameters you set for them.

APPENDIX A

ACTIVITIES & RESOURCES

Stylized Numbers 80
5-Frame Activities 100
Dot Card Activities 107
Extended 5-Frames 122
Mini-books to Make 124
Felt House Pattern 126
Number Families 130
Practice Houses by Street 134
Subtraction Houses 139
Mixed Practice Houses 143
Write Problems 153
Mixed Practice, Houses and Problems 159
Blank Practice Houses, Full Sheet 165
Word Problems 166

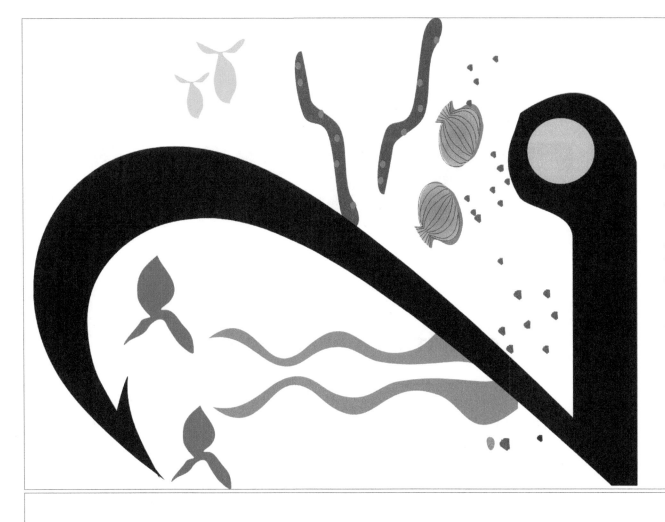

2 is a fishing hook ready for bait.

1 is a thin man standing up straight,

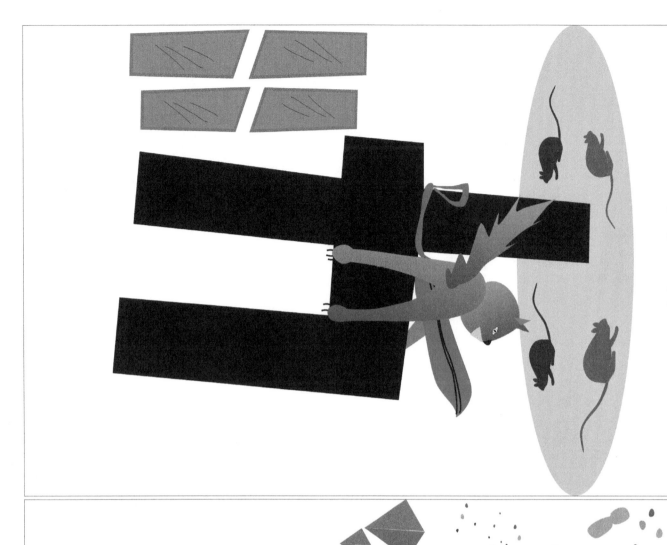

4 is a chair that is turned around.

3 is a camel sitting on the ground;

6 is a unicycle, watch me ride!

5 has a tummy too big to hide;

8 is a snowman, makes me laugh!

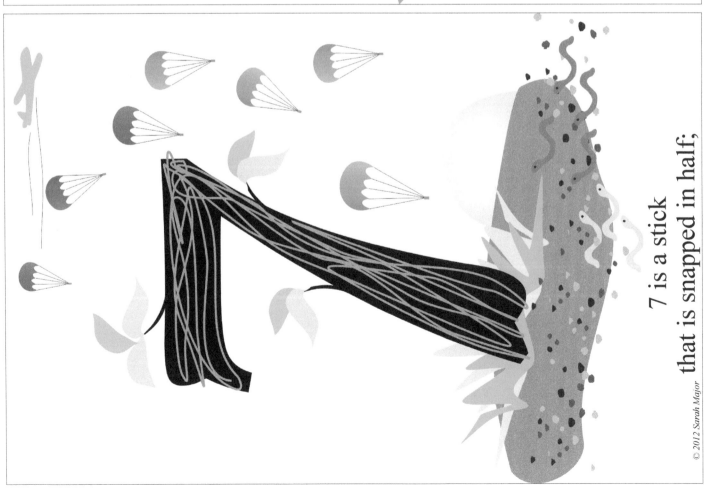

7 is a stick that is snapped in half;

10 is a thin man, hear him say,
"Better watch out or you'll roll away!"

9 is a red balloon floating up high;
if I let go, I think I'll cry.

12 likes to dig and delve.

McNeven twins are 11.

14's chair is in between.

13 has a green canteen.

16 is a tall marine.

15's keys are really clean.

18 likes some evergreens.

17 will cut the beans.

20 always eats a-plenty!

19 sees a small Colleen.

1	2	3	4	5
6	7	8	9	10

4.3. Five Frame with Numbers to 10   **Name:**_____

1	2	3	4	5
6	7	8	9	10

4.3. Five Frame with Numbers to 10   **Name:**_____

1	2	3	4	5
6	7	8	9	10

4.3. Five Frame with Numbers to 10   **Name:**_____

4.4a. Ordering Numbers to 10 Name:_____

3   6   1   8   2   4   7   9   5   10

- -

4.4a. Ordering Numbers to 10 Name:_____

3   6   1   8   2   4   7   9   5   10

4.4b. Ordering Numbers to 20   **Name:**_____

3   6   1   8   2

4   7   9   5   10

12   15   11   16   18

20   13   17   19   14

	2			5
6		8		10

4.5a. Missing Numbers to Ten   **Name:**_____

		3		5
6	7		9	

4.5a. Missing Numbers to Ten   **Name:**_____

1	2		4	5
6		8		

4.5a. Missing Numbers to Ten   **Name:**_____

4.5b. Missing Numbers to 20   Name:_____

1		3	4	
6			9	
	12	13		15
		18		20

- -

4.5b. Missing Numbers to 20   Name:_____

	2			5
	7	8		10
11			14	
16	17		19	

104

4.6a. Five Frame Chart to 20   Name:_____

1	2	3	4	5
6	7	8	9	10
11	12	13	14	15
16	17	18	19	20

4.6b. Blank Five Frame Chart for Practice   Name:_____

4.7. Blank Five-Frame to Complete Name:_____

4.7. Blank Five-Frame to Complete Name:_____

4.7. Blank Five-Frame to Complete Name:_____

5.1. Blank Dot Grid

5.2a. Dot Cards, Page 1

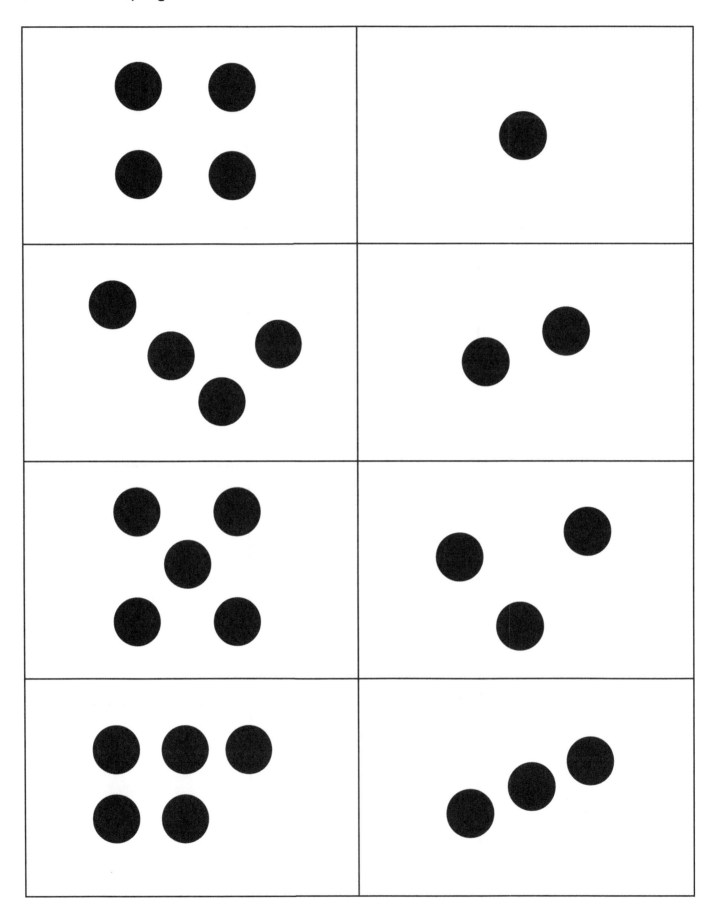

5.2b. Dot Cards, Page 2

BINGO

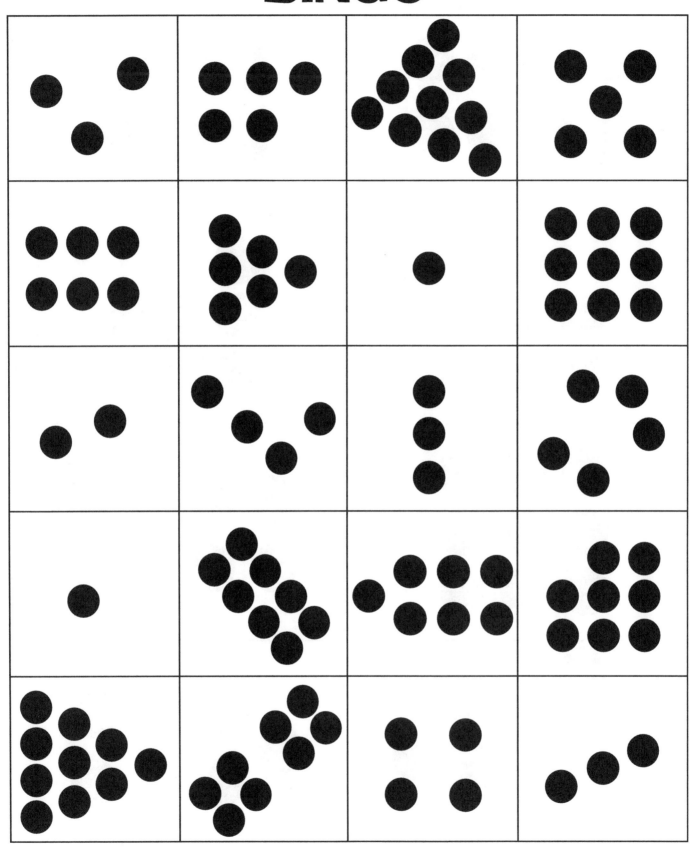

5.3a. Dot Card Bingo Board 1

BINGO

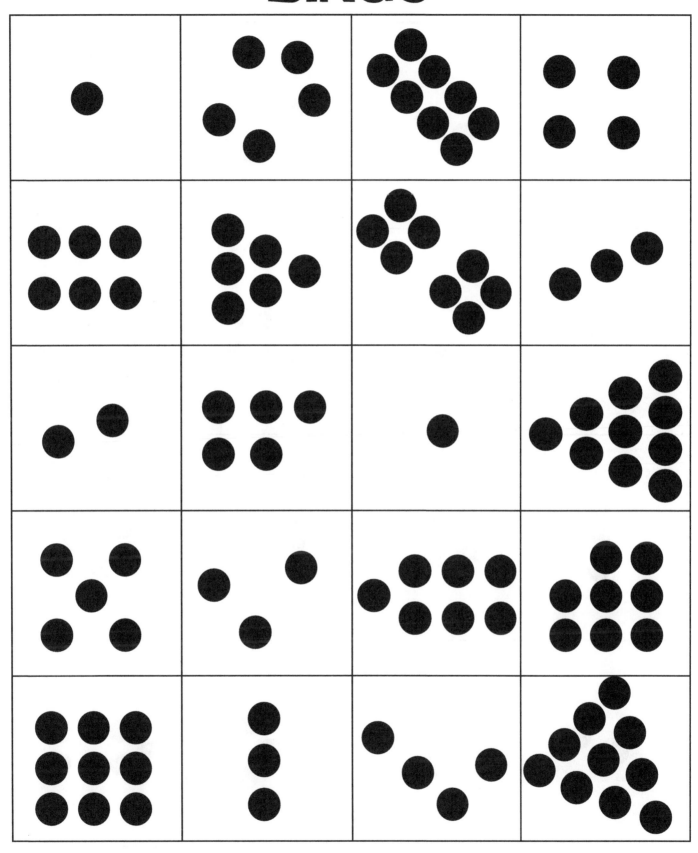

5.3b. Dot Card Bingo Board 2

5.4a. Floor Dominoes, Page 1.

5.4b. Floor Dominoes, Page 2.

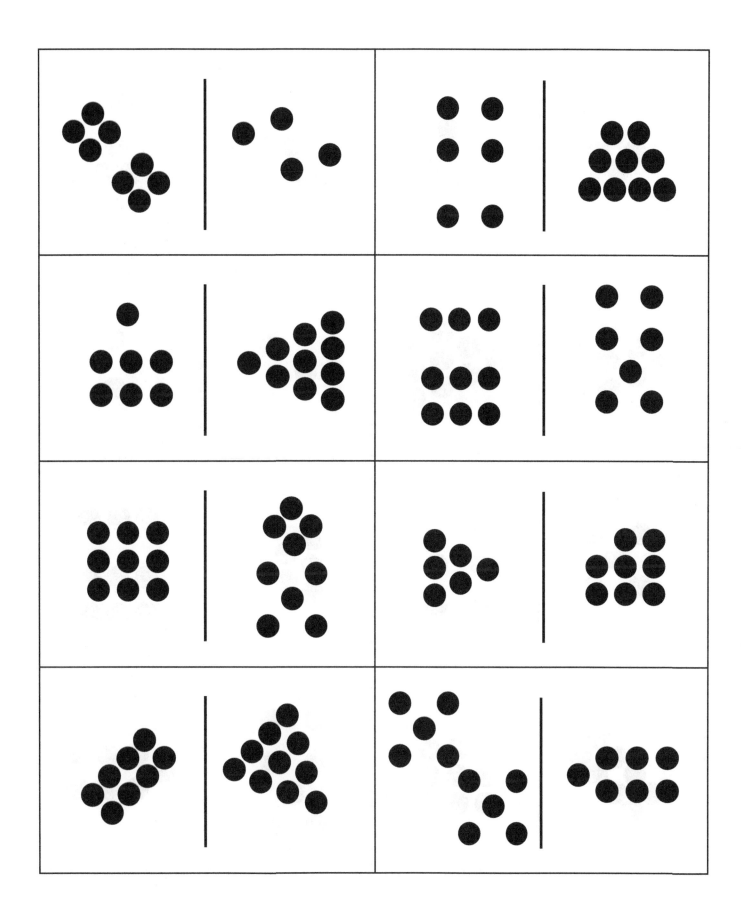

5.4c. Floor Dominoes, Page 3.

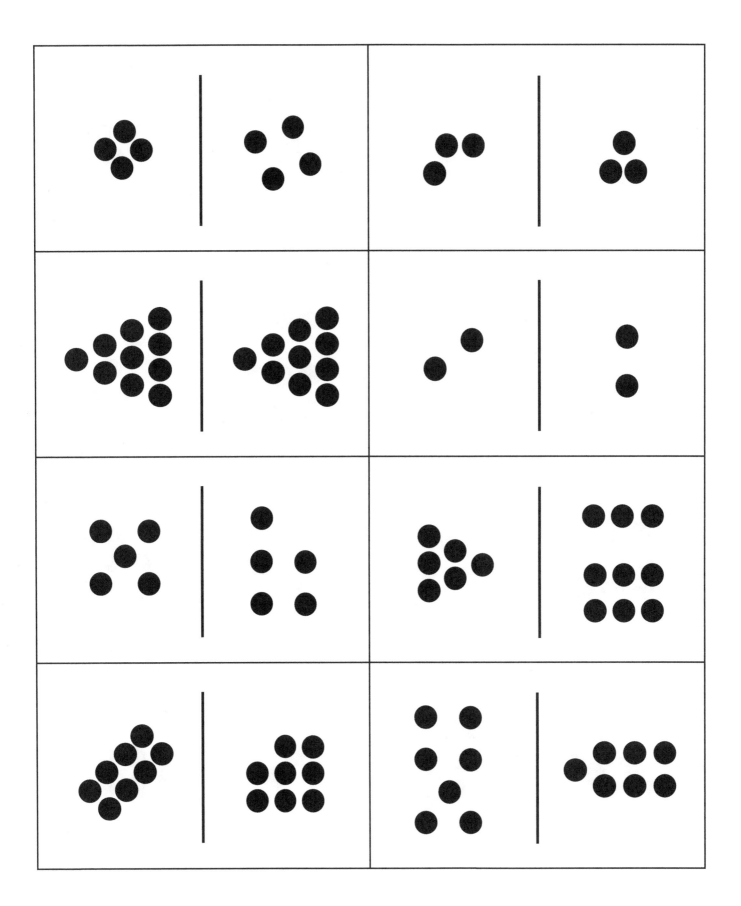

5.4d. Floor Dominoes, Page 4.

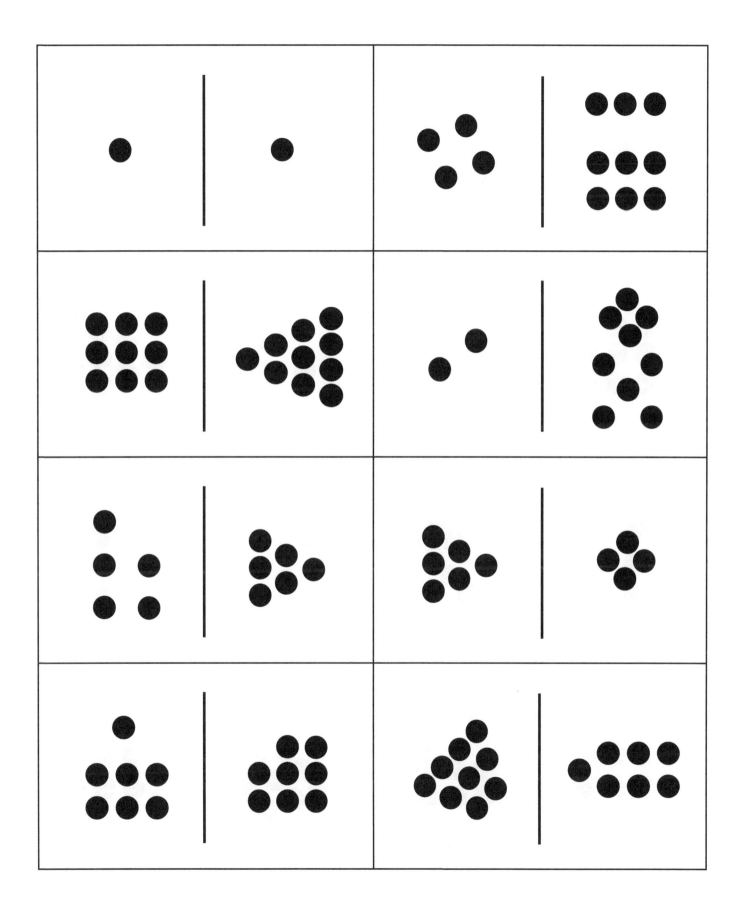

5.4e. Floor Dominoes, Page 5.

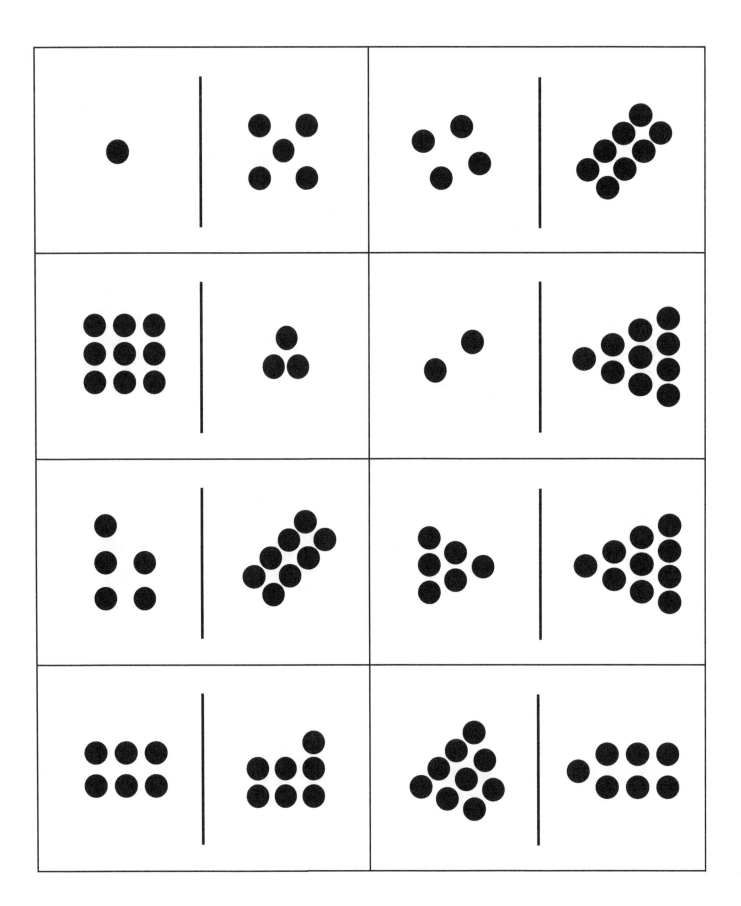

5.4f. Floor Dominoes, Page 6.

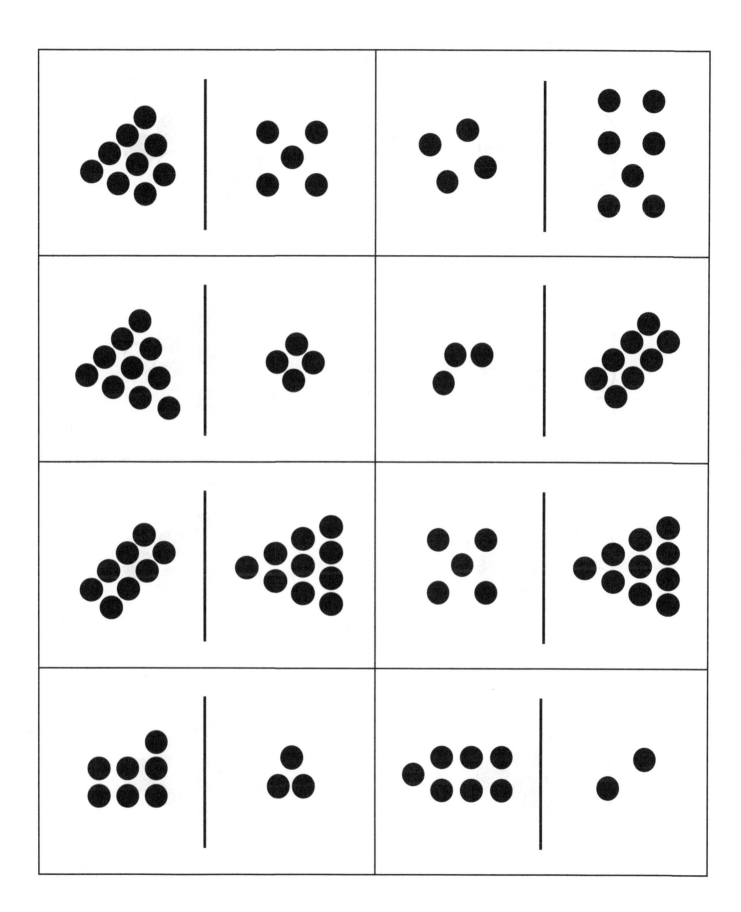

5.5a. Pyramid Dot Cards, Page 1.

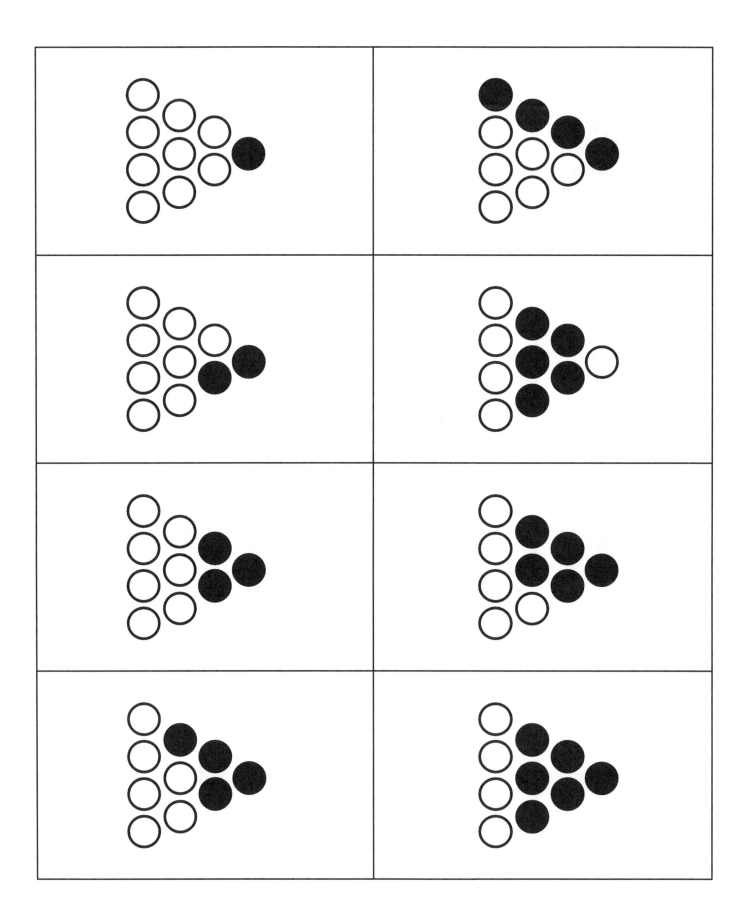

5.5b. Pyramid Dot Cards, Page 2.

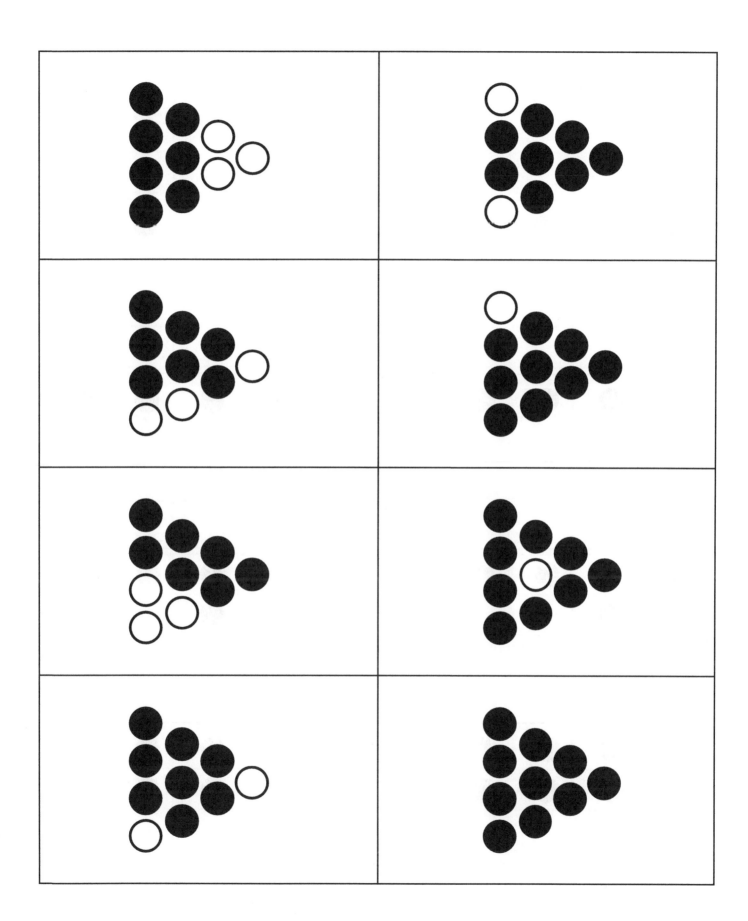

5.6a. 5-Frame Dot Cards, Page 1.

5.6b. 5-Frame Dot Cards, Page 2.

5.7. 5-Frame Cards to 70.

1	2	3	4	5
6	7	8	9	10
11	12	13	14	15
16	17	18	19	20
21	22	23	24	25
26	27	28	29	30
31	32	33	34	35
36	37	38	39	40
41	42	43	44	45
46	47	48	49	50
51	52	53	54	55
56	57	58	59	60
61	62	63	64	65
66	67	68	69	70

1	2	3	4	5
6	7	8	9	10
11	12	13	14	15
16	17	18	19	20
21	22	23	24	25
26	27	28	29	30
31	32	33	34	35
36	37	38	39	40
41	42	43	44	45
46	47	48	49	50
51	52	53	54	55
56	57	58	59	60
61	62	63	64	65
66	67	68	69	70

5.8. Blank 5-Frame Cards to 70.

When You Say....
I See....

by _____

My Book of Numbers

When You Say....
I See....

by _____

My Book of Numbers

My Dot Critter Book

by _____

My Second Book of Numbers

My Dot Critter Book

by _____

My Second Book of Numbers

7.1. Felt-House Pattern

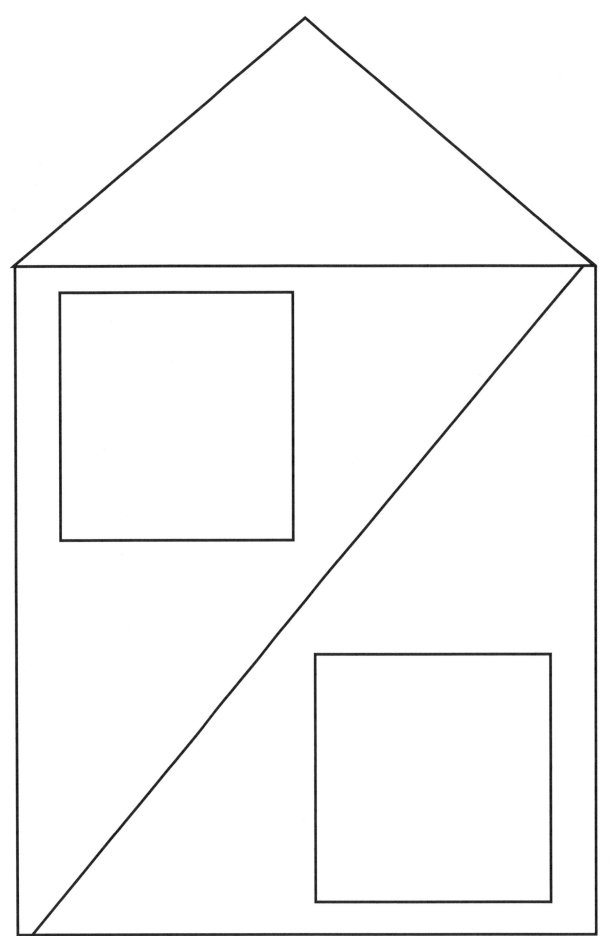

7.2a. Attic Numbers, Page 1

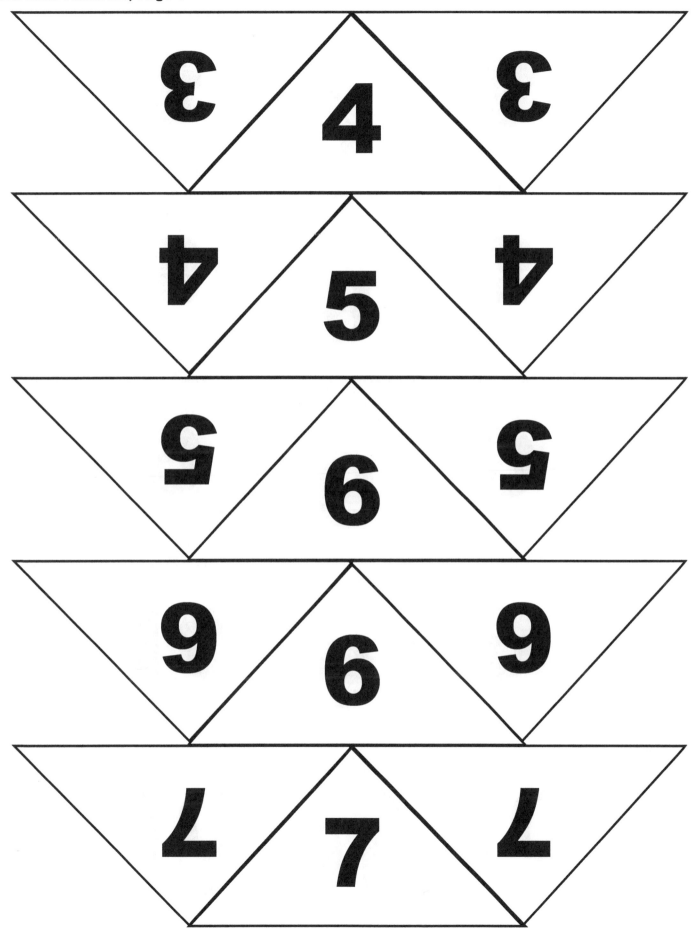

7.2b. Attic Numbers, Page 2

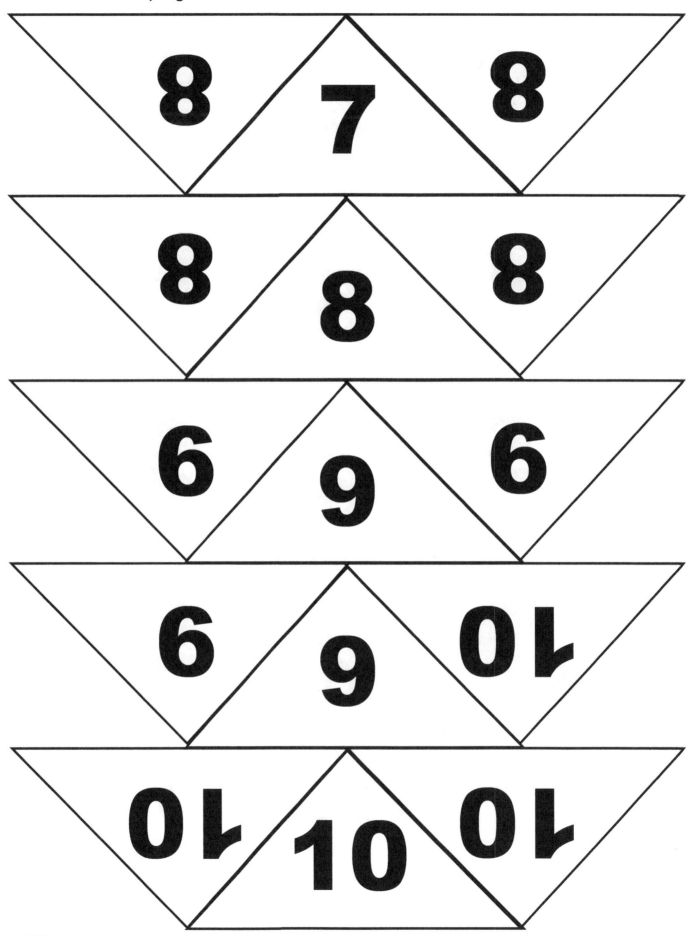

7.3. Window Numbers, and Attic Numbers, cont.

| 10 | 10 |

0	1	2	2
3	3	4	4
5	5	6	7
8	9	10	

7.4a. Threes
Figure out which families can live in each house.

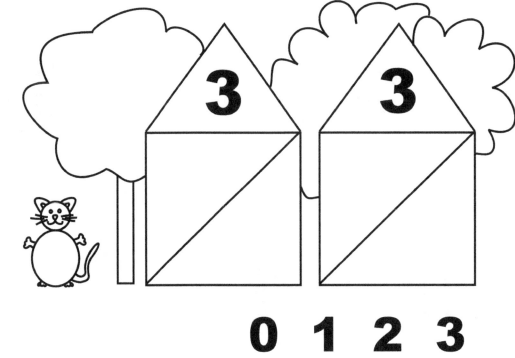

Name:_____

7.4b. Fours
Figure out which families can live in each house.

Name:_____

7.4c. Fives
Figure out which families can live in each house.

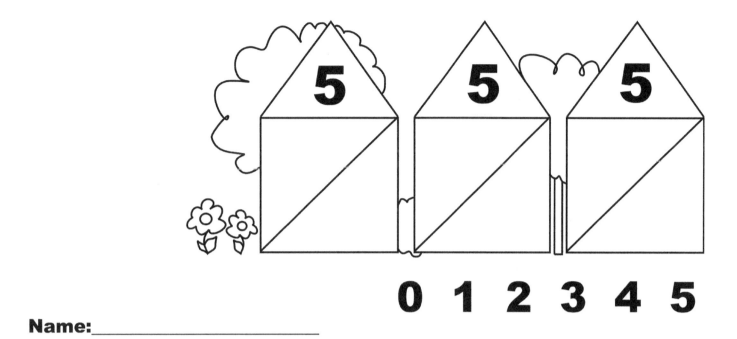

Name:_____

7.4d. Sixes
Figure out which families can live in each house.

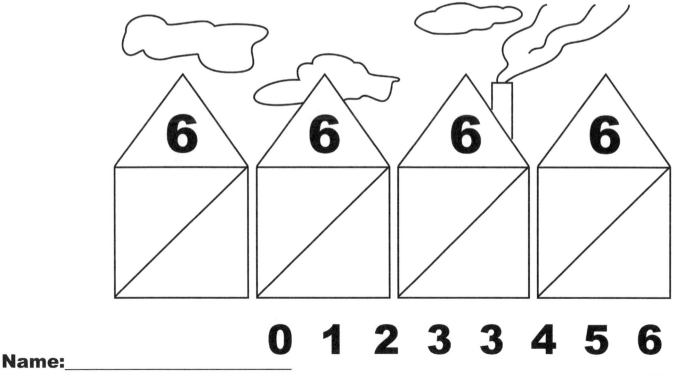

Name:_____

7.4e. Sevens

Figure out which families can live in each house.

0 1 2 3 4 5 6 7

Name:_____

7.4f. Eights

Figure out which families can live in each house.

0 1 2 3 4 4 5 6 7 8

Name:_____

7.4g. Nines
Figure out which families can live in each house.

Name:_____

7.4h. Tens
Figure out which families can live in each house.

Name:_____

Name_____

fold

fold

7.5a. Third Street practice houses

Name_____

fold

fold

7.5b. Fourth and Fifth Street practice houses

Name_____

fold

fold

7.5c. Sixth and Seventh Street practice houses

Name_____

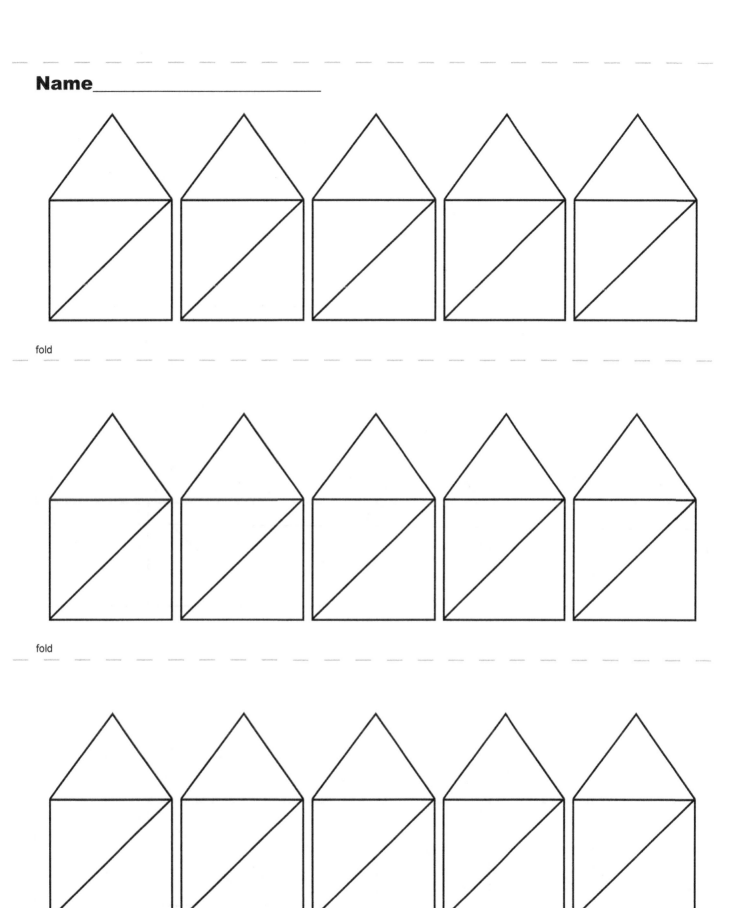

fold

fold

7.5d. Eighth and Ninth Street practice houses

Name _____

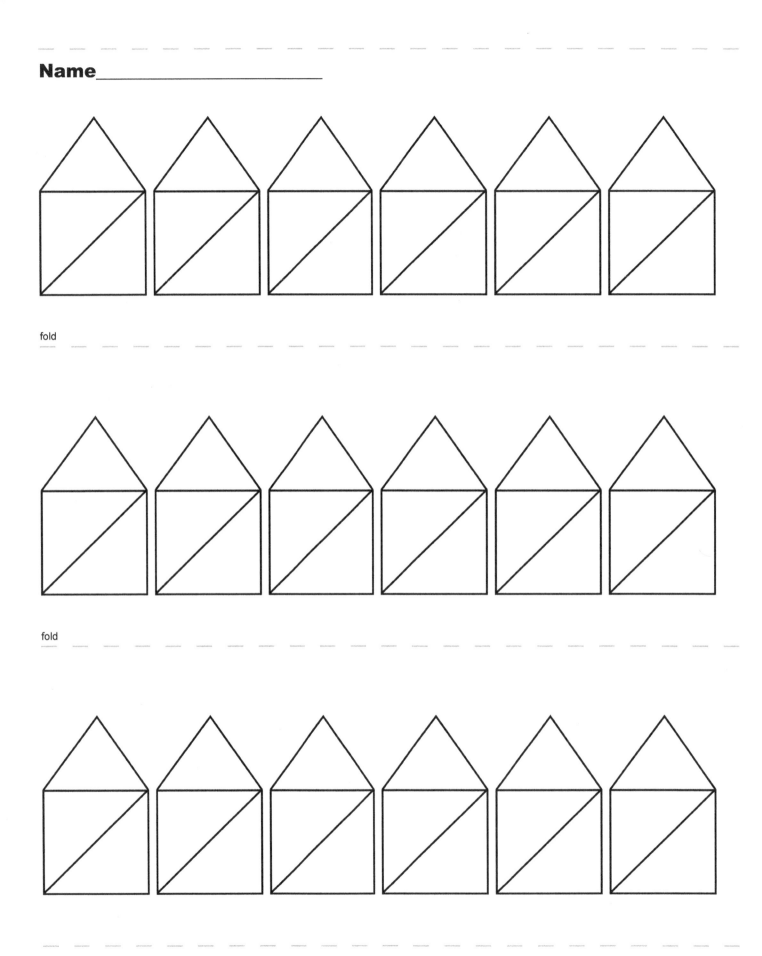

fold

fold

7.5e. Tenth Street practice houses

7.6a. Subtraction problems for three

Please help me figure out which families went walking.

Name:_____

7.6b. Subtraction Problems for four

Please help me figure out which families went walking.

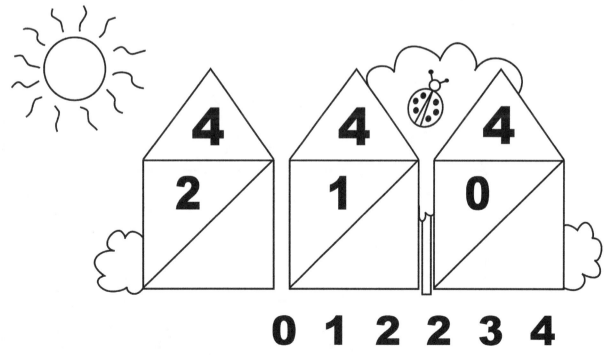

Name:_____

7.6c. Subtraction problems for five

Please help me figure out which families went walking.

Name:_____

7.6d. Subtraction problems for six

Please help me figure out which families went walking.

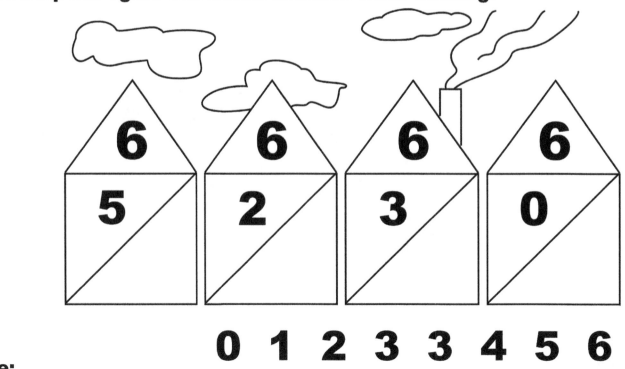

Name:_____

7.6e. Subtraction problems for seven

Please help me figure out which families went walking.

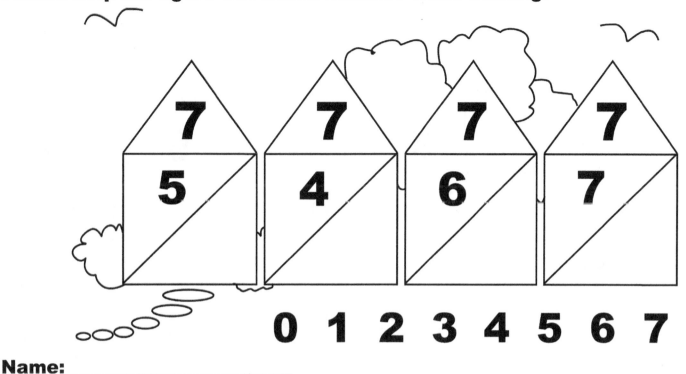

Name:_____

7.6f. Subtraction problems for eight

Please help me figure out which families went walking.

Name:_____

7.6g. Subtraction problems for nine

Please help me figure out which families went walking.

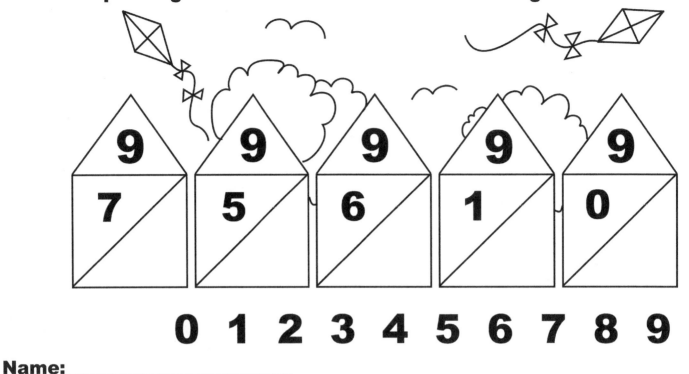

Name:_____

7.6h. Subtraction problems for ten

Please help me figure out which families went walking.

Name:_____

7.7a. Practice for threes   Name: _____

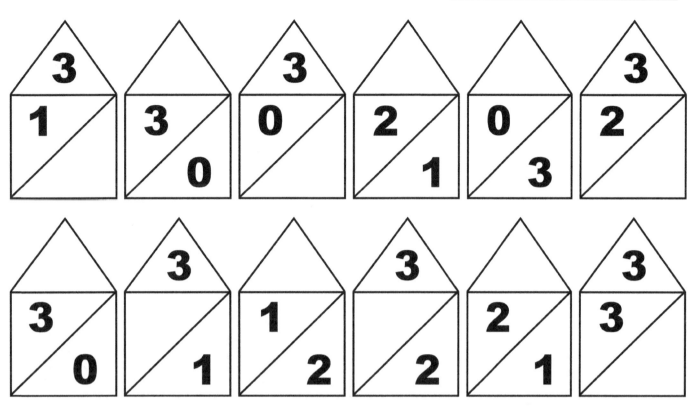

7.7b. Practice for fours   Name: _____

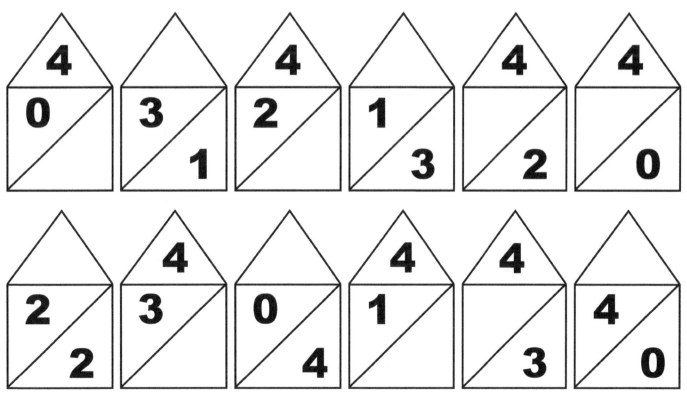

7.7c. Practice for fives   **Name:** _____

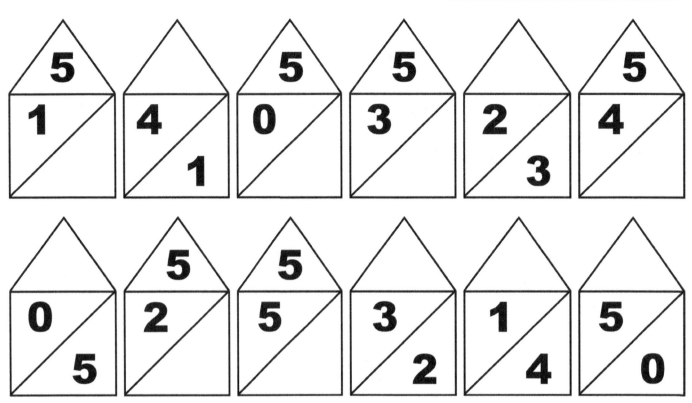

7.7d. Practice for sixes   **Name:** _____

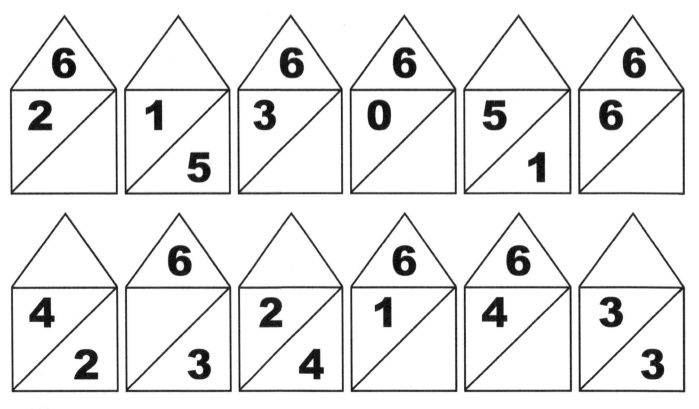

7.7e. Practice for sevens   Name: _____

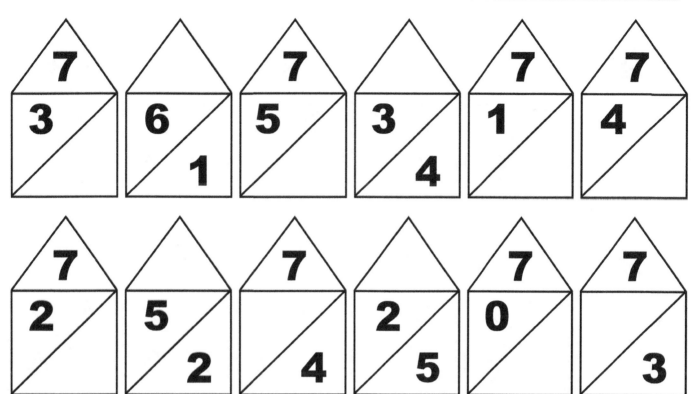

7.7f. Practice for eights   Name: _____

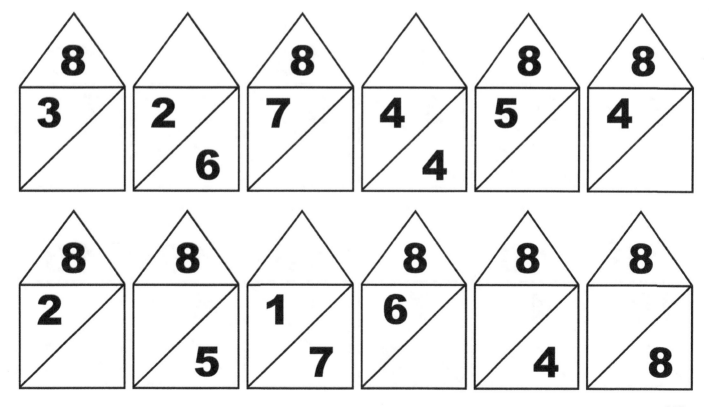

145

7.7g. Practice for nines   **Name:** _____

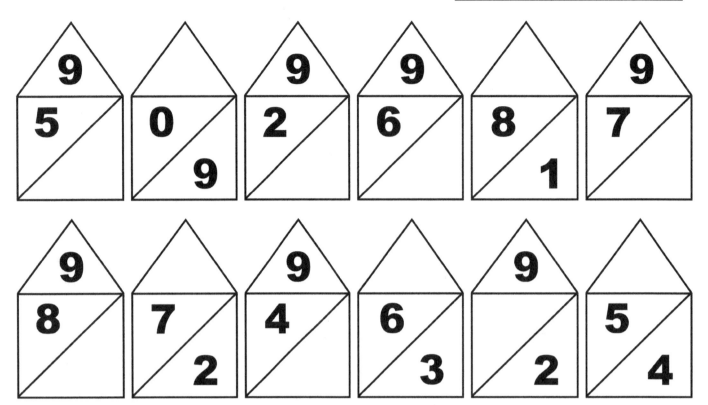

7.7h. Practice for tens   **Name:** _____

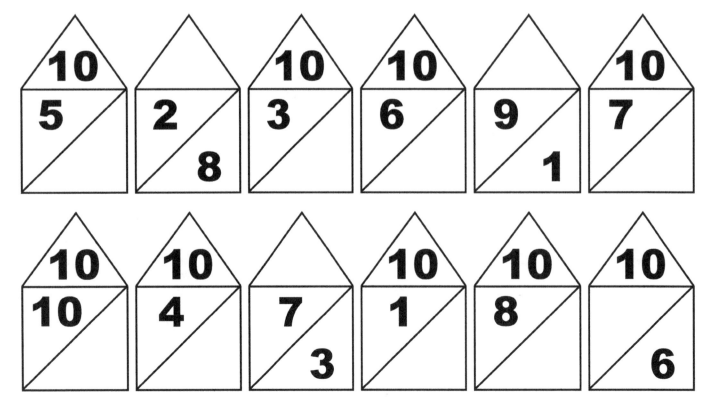

7.8a. From houses to number problems - threes   Name: _____

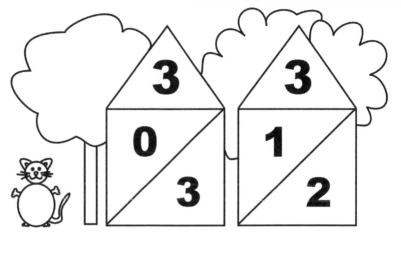

2	3	3	1	3	3	3	0
+1	-2	+0	+2	-1	-0	-3	+3

Use the houses above to solve each problem.

7.8b. From houses to number problems - fours   Name: _____

1
+ 3

4
- 3

4	4	0	2	4	4	4	3
-2	-0	+4	+2	-1	+0	-4	+1

Use the houses above to solve each problem.

7.8c. From houses to number problems - fives   **Name:** _____

2	5	1	5	5	5	5	5
+3	+0	+4	-4	-2	-5	-1	-3

Use the houses above to solve each problem.

7.8d. From houses to number problems - sixes   **Name:** _____

2
+4

6
+0

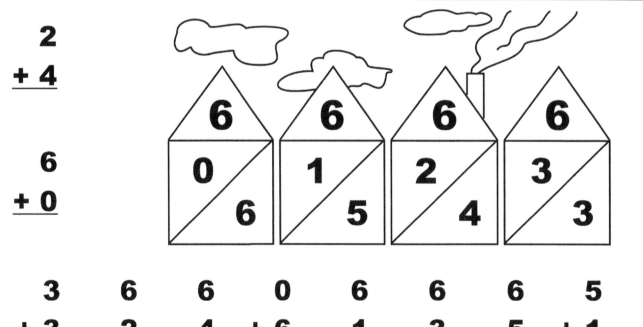

3	6	6	0	6	6	6	5
+3	-2	-4	+6	-1	-3	-5	+1

Use the houses above to solve each problem.

7.8e. From houses to number problems - sevens   Name: _____

Use the houses below to solve each problem.

```
  7     3     0     7     2     6     7     5
 -3    +4    +7    -2    +5    +1    -5    +2
 ___   ___   ___   ___   ___   ___   ___   ___
```

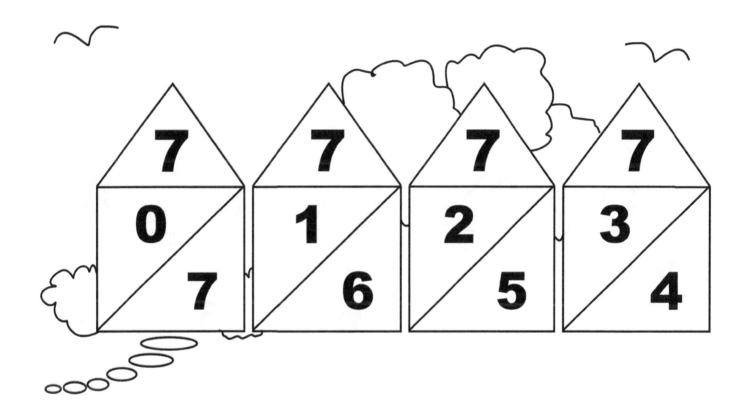

```
  7     7     1     7     7     7     7     4
 -0    -6    +6    -1    -4    -7    +0    +3
 ___   ___   ___   ___   ___   ___   ___   ___
```

7.8f. From houses to number problems - eights   Name: _____

Use the houses below to solve each problem.

```
  0     3     8     6     8     7     8     4
+ 8   + 5   - 4   + 2   - 1   + 1   - 2   + 4
```

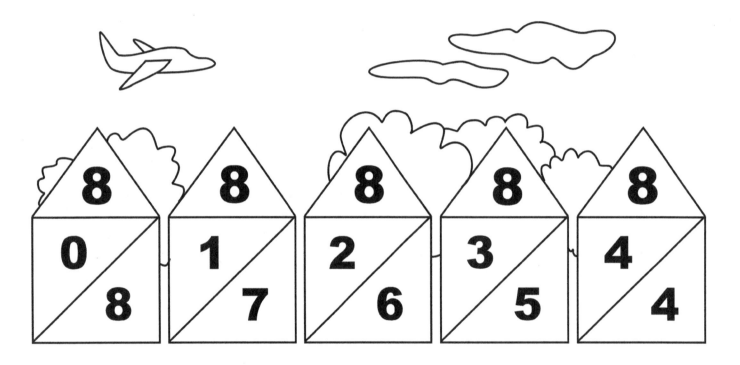

```
  8     8     5     2     8     8     1     8
- 6   - 5   + 3   + 6   - 7   - 3   + 7   - 0
```

150

7.8g. From houses to number problems - nines   Name: _____

Use the houses below to solve each problem.

```
  9     9     9     1     9     9     0     8
- 8   - 1   - 4   + 8   - 0   - 1   + 9   + 1
___   ___   ___   ___   ___   ___   ___   ___
```

```
  2     9     7     6     9     4     2     9
+ 7   - 7   + 2   + 3   - 3   + 5   + 7   - 2
___   ___   ___   ___   ___   ___   ___   ___
```

151

7.8h. From houses to number problems - tens   Name: _____

Use the houses below to solve each problem.

```
 10    10    10     9    10    10    10     8
- 6   - 4   - 1   + 1   - 5   - 2   + 0   + 2
___   ___   ___   ___   ___   ___   ___   ___
```

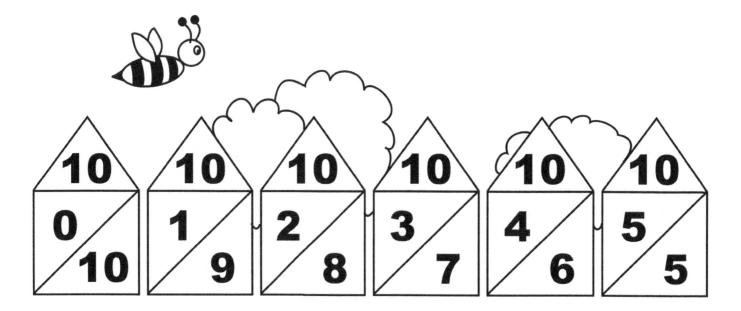

```
 10     3     8     5    10     7     4     2
- 3   + 7   + 2   + 5   - 3   + 3   + 6   + 8
___   ___   ___   ___   ___   ___   ___   ___
```

7.9a. Writing problems - threes

Name: _____

Write problems for each house.

- -

7.9b. Writing problems - fours

Name: _____

Write problems for each house.

7.9c. Writing problems - fives

Name: _____

Write problems for each house.

7.9d. Writing problems - sixes

Name: _____

Write problems for each house.

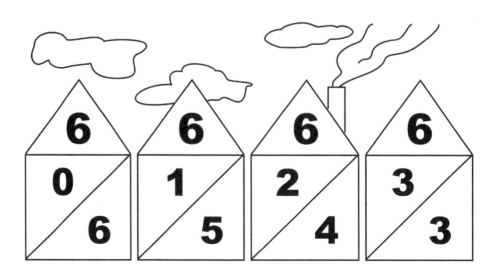

154

7.9e. Writing problems - sevens

Name: _____

Write problems for each house.

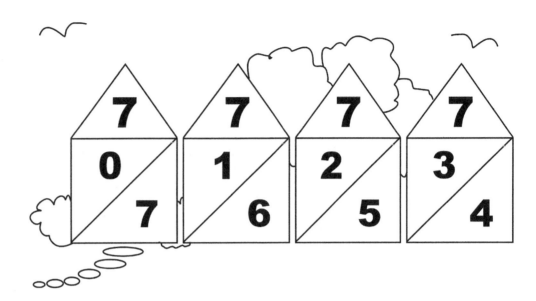

7.9f. Writing problems - eights

Name: _____

Write problems for each house.

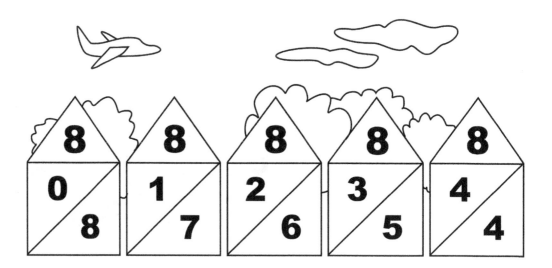

7.9g. Writing problems - nines

Name: _____

Write problems for each house.

157

7.9h. Writing problems - tens

Name: _____

Write problems for each house.

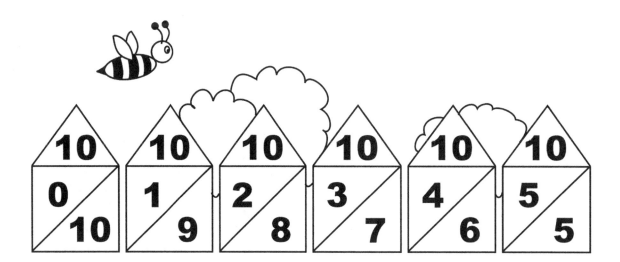

158

7.10a. Addition and subtraction practice - threes and fours

Name: _____

```
  3     4     1     2     3     0     4     1
- 0   - 4   + 2   + 2   - 1   + 3   - 2   + 3
```

```
  4     3     3     4     4     3     3     2
- 3   - 2   + 1   + 0   - 1   + 0   - 3   + 1
```

    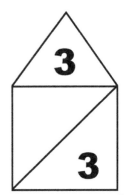

7.10b. Addition and subtraction practice - fives and sixes

Name: _____

    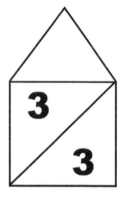

```
  6      5      0      3      5      4      5      1
- 5    - 3    + 5    + 3    - 1    + 2    - 0    + 4
```

```
  6      5      3      6      6      5      6      6
- 4    - 5    + 2    + 0    - 2    + 1    - 3    - 1
```

7.10c. Addition and subtraction practice - threes, fours, fives and sixes

Name: _____

Row 1: top 5 / 2, _ ; top _ / 2, 4 ; top 6 / 3, _ ; top _ / 1, 4 ; top _ / 2, 1

Row 2: top _ / 3, 3 ; top 5 / 4, _ ; top _ / 3, 2 ; top 4 / 1, _ ; top 6 / _, 5

Row 3: top 4 / _, 2 ; top 3 / 1, _ ; top 4 / _, 3 ; top _ / 0, 5 ; top 3 / 3, _

Row 4: top 6 / 2, _ ; top _ / 1, 3 ; top 4 / 0, _ ; top 3 / _, 2 ; top 5 / 1, _

Row 5: top _ / 0, 3 ; top 5 / _, 3 ; top _ / 1, 5 ; top 6 / 4, _ ; top _ / 2, 2

161

7.10d. Addition and subtraction practice - sevens and eights

Name: _____

8	7	8	5	7	3	8	8
- 4	- 5	+ 0	+ 3	- 6	+ 4	- 2	- 8

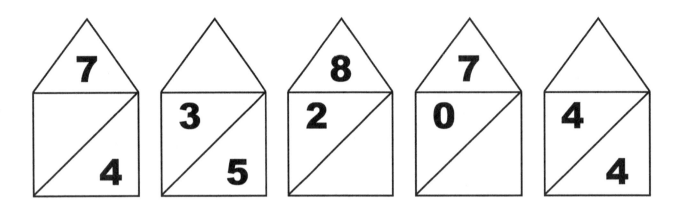

7	8	1	4	7	5	7	1
- 3	- 1	+ 6	+ 4	- 0	+ 2	- 1	+ 7

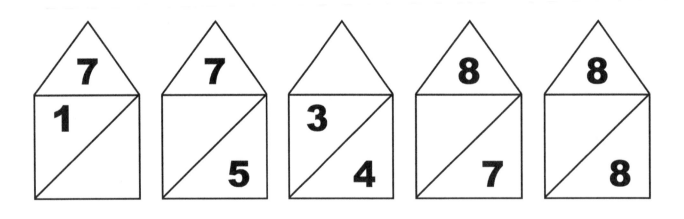

8	8	0	2	7	4	7	8
- 3	- 7	+ 7	+ 6	- 2	+ 3	- 4	- 6

7.10e. Addition and subtraction practice - nines and tens

Name: _____

```
 10    9    10    6    9    1    10    9
- 4   - 5   + 0   + 3  - 6   + 8  - 2   - 9
```

```
 10    10    7    5    10    3    9    8
- 7   - 1   + 2   + 5  - 0   + 7  - 1   + 2
```

```
 9    10    0    4    9    4    9    10
- 2   - 8   + 9   + 6  - 3   + 5  - 8   - 5
```

7.10f. Addition and subtraction practice - sevens, eights, nines and tens

Name: _____

7.11 Extra practice houses

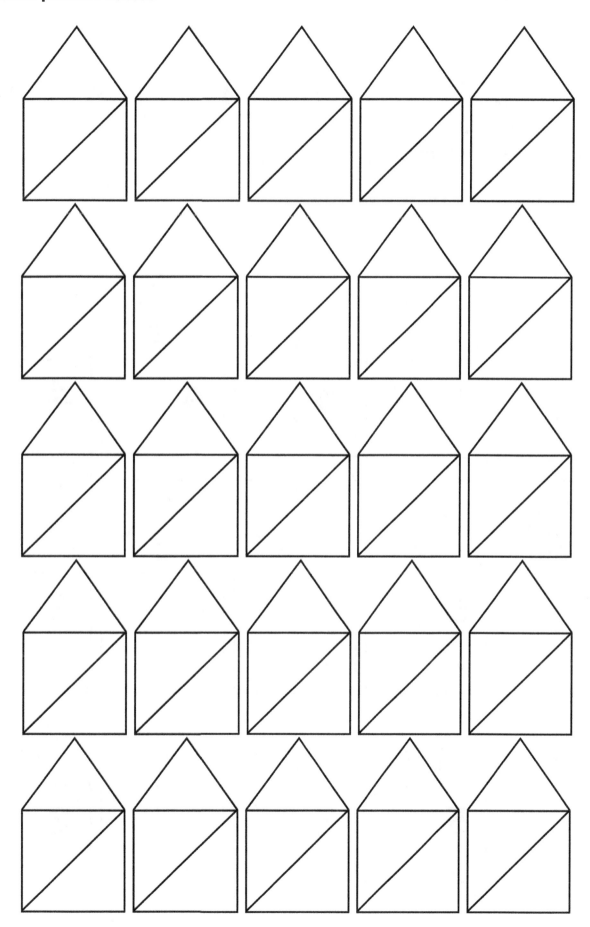

7.12a. Mixed subtraction word problem

Name: _____

One family from each house went to a play. You are the bus driver. You must take each family to the right house. Use the numbers in the bus as many times as you need to.

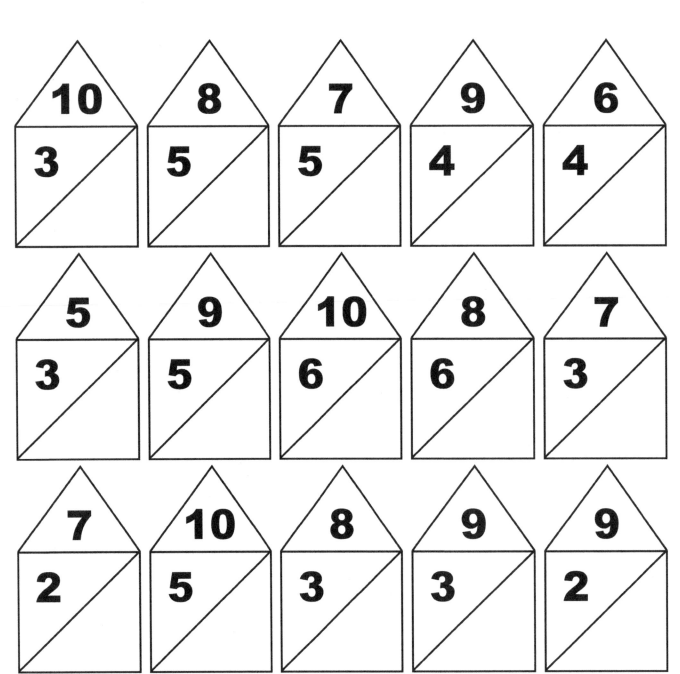

166

7.12b. Mixed addition word problem

Name: _____

You are the new soccer coach. You need to choose which teams play each other. Place players on the field. Make sure there are the same number of players on each side. The first problem is done for you.

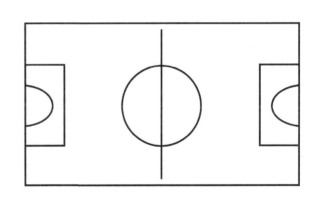

Choose from these teams:

6	2	5	5	6	5	2	7	8	8
+4	+7	+5	+4	+3	+3	+6	+3	+1	+2

167

7.12c. Mixed subtraction word problem

Name: _____

It was the big Summer Celebration. But right at the start, a big wind began to blow. Each family at the picnic lost 3 paper plates. How many did each family have left?

$$4 - 3$$

$$3 - 3$$

$$7 - 3$$

$$5 - 3$$

$$8 - 3$$

$$9 - 3$$

$$10 - 3$$

$$6 - 3$$

7.12d. Draw your own problems

Name: _____

Choose a street you want to draw. Draw the houses and fill in the families that live in each house. Decorate the street any way you wish.

7.12e. Draw your own problems

Name: _____

Make up a story about any street. Draw numbers in the houses to show what happens in your story.

APPENDIX B

MONITORING & TRACKING FORMS

Clipboard Assessment for Learning Numbers 172
Progress Report (from Chapter 4) 173
Whole Class Record (from Chapters 5-7) 174
Whole Class Record (from Chapter 7) 175
Individual Assessment: Computation to Ten 176

Clipboard Assessment for Learning Numbers (Chapter 4)

Name	Skill: 1 Recognize numbers to ten		2 Count to ten		3 Order numbers sequentially		4 Supply missing numbers		5 Discover patterns I = interested V = very interested T = taking it further	6 Write numbers	
	1-10	11-20	1-10	11-20	1-10	11-20	1-10	11-20		1-10	11-20

Progress Report (Chapter 4)

_____'s Numbers   Date: _____

	10	20	Other
1. I can recognize numbers to....................			
2. I can count to..			
3. I can order numbers to..........................			
4. I can supply missing numbers to............			
5. I can write numbers to..........................			

Teacher's comments:

Progress Report (Chapter 4)

_____'s Numbers   Date: _____

	10	20	Other
1. I can recognize numbers to....................			
2. I can count to..			
3. I can order numbers to..........................			
4. I can supply missing numbers to............			
5. I can write numbers to..........................			

Teacher's comments:

Whole Class Record (Chapters 5-7)

Check off or date each skill when student shows mastery

Name	Dot Cards	Five Frames to 20	Five Frames to 70	My Two Hands	Houses

Whole Class Record (Chapter 7)

Check off or date each street when student shows mastery

Name	Third	Fourth	Fifth	Sixth	Seventh	Eighth	Ninth	Tenth

Right-Brained Addition & Subtraction, 2nd Edition, © 2013 Sarah Major, Child1st Publications, www.child1st.com.

Individual Assessment: Computation to Ten (Chapter 7)

Name: _____

Based on several observations and samples of written work, draw a line through each problem as it is mastered. When an entire street has been mastered, enter the date of mastery. Note any needs for additional practice in the last column.

			Street				Date of Mastery	Extra Practice Needed
Third:	0+3	1+2						
Fourth:	0+4	1+3	2+2					
Fifth:	0+5	1+4	2+3					
Sixth:	0+6	1+5	2+4	3+3				
Seventh:	0+7	1+6	2+5	3+4				
Eighth:	0+8	1+7	2+6	3+5	4+4			
Ninth:	0+9	1+8	2+7	3+6	4+5			
Tenth:	0+10	1+9	2+8	3+7	4+6	5+5		

Individual Assessment: Computation to Ten (Chapter 7)

Name: _____

Based on several observations and samples of written work, draw a line through each problem as it is mastered. When an entire street has been mastered, enter the date of mastery. Note any needs for additional practice in the last column.

			Street				Date of Mastery	Extra Practice Needed
Third:	0+3	1+2						
Fourth:	0+4	1+3	2+2					
Fifth:	0+5	1+4	2+3					
Sixth:	0+6	1+5	2+4	3+3				
Seventh:	0+7	1+6	2+5	3+4				
Eighth:	0+8	1+7	2+6	3+5	4+4			
Ninth:	0+9	1+8	2+7	3+6	4+5			
Tenth:	0+10	1+9	2+8	3+7	4+6	5+5		

APPENDIX C

ANSWER KEYS

130.
- **7.4a.** 0 / 3 or 3 / 0
  1 / 2 or 2 / 1
- **7.4b.** 0 / 4 or 4 / 0
  1 / 3 or 3 / 1
  2 / 2

131.
- **7.4c.** 0 / 5 or 5 / 0
  1 / 4 or 4 / 1
  2 / 3 or 3 / 2
- **7.4d.** 0 / 6 or 6 / 0
  1 / 5 or 5 / 1
  2 / 4 or 4 / 2
  3 / 3

132.
- **7.4e.** 0 / 7 or 7 / 0
  1 / 6 or 6 / 1
  2 / 5 or 5 / 2
  3 / 4 or 4 / 3
- **7.4f.** 0 / 8 or 8 / 0
  1 / 7 or 7 / 1
  2 / 6 or 6 / 2
  3 / 5 or 5 / 3
  4 / 4

133.
- **7.4g.** 0 / 9 or 9 / 0
  1 / 8 or 8 / 1
  2 / 7 or 7 / 2
  3 / 6 or 6 / 3
  4 / 5 or 5 / 4
- **7.4h.** 0 / 10 or 10 / 0
  1 / 9 or 9 / 1
  2 / 8 or 8 / 2
  3 / 7 or 7 / 3
  4 / 6 or 6 / 4
  5 / 5

139.
- **7.6a.** 1, 3
- **7.6b.** 2, 3, 4

140.
- **7.6c.** 5, 2, 1
- **7.6d.** 1, 4, 3, 6

141.
- **7.6e.** 2, 3, 1, 0
- **7.6f.** 1, 3, 6, 4, 8

142.
- **7.6g.** 2, 4, 3, 8, 9
- **7.6h.** 1, 10, 4, 5, 7, 8

143.
- **7.7a.** 2, 3, 3, 3, 3, 1
  3, 2, 3, 1, 3, 0
- **7.7b.** 4, 4, 2, 4, 2, 4
  4, 1, 4, 3, 1, 4

144.
- **7.7c.** 4, 5, 5, 2, 5, 1
  5, 3, 0, 5, 5, 5
- **7.7d.** 4, 6, 3, 6, 6, 0
  6, 3, 6, 5, 2, 6

145.
- **7.7e.** 4, 7, 2, 7, 6, 3
  5, 7, 3, 7, 7, 4
- **7.7f.** 5, 8, 1, 8, 3, 4
  6, 3, 8, 2, 4, 0

146.
- **7.7g.** 4, 9, 7, 3, 9, 2
  1, 9, 5, 9, 7, 9
- **7.7h.** 5, 10, 7, 4, 10, 3
  0, 6, 10, 9, 2, 4

147.
- **7.8a.** 3, 1, 3, 3, 2, 3, 0, 3
- **7.8b.** 4, 1, 2, 4, 4, 4, 3, 4, 0, 4

148.
- **7.8c.** 5, 5, 5, 1, 3, 0, 4, 2
- **7.8d.** 6, 6, 6, 4, 2, 6, 5, 3, 1, 6

149.
- **7.8e.** 4, 7, 7, 5, 7, 7, 2, 7
  7, 1, 7, 6, 3, 0, 7, 7

150.
- **7.8f.** 8, 8, 4, 8, 7, 8, 6, 8
  2, 3, 8, 8, 1, 5, 8, 8

151.
- **7.8g.** 1, 8, 5, 9, 9, 8, 9, 9
  9, 2, 9, 9, 6, 9, 9, 7

152.
- **7.8h.** 4, 6, 9, 10, 5, 8, 10, 10
  7, 10, 10, 10, 7, 10, 10, 10

Please Note:
Large number is page number; medium numbers identify activities.
In answers 7.4a-h, / separates the upstairs from the downstairs.
In answers separated by commas, each row of answers represents one row of problems.
We have provided all possible answers for problems 7.9a-h.

153.

7.9a. House 1:
3 - 0 = 3
3 - 3 = 0
3 + 0 = 3
0 + 3 = 3

House 2:
3 - 1 = 2
3 - 2 = 1
2 + 1 = 3
1 + 2 = 3

7.9b. House 1:
4 - 0 = 4
4 - 4 = 0
4 + 0 = 4
0 + 4 = 4

House 2:
4 - 1 = 3
4 - 3 = 1
3 + 1 = 4
1 + 3 = 4

House 3:
4 - 2 = 2
2 + 2 = 4

154.

7.9c. House 1:
5 - 0 = 5
5 - 5 = 0
5 + 0 = 5
0 + 5 = 5

House 2:
5 - 1 = 4
5 - 4 = 1
4 + 1 = 5
1 + 4 = 5

House 3:
5 - 2 = 3
5 - 3 = 2
3 + 2 = 5
2 + 3 = 5

7.9d. House 1:
6 - 0 = 6
6 - 6 = 0
6 + 0 = 6
0 + 6 = 6

House 2:
6 - 1 = 5
6 - 5 = 1
5 + 1 = 6
1 + 5 = 6

House 3:
6 - 2 = 4
6 - 4 = 2
4 + 2 = 6
2 + 4 = 6

House 4:
6 - 3 = 3
3 + 3 = 6

155.

7.9e. House 1:
7 - 0 = 7
7 - 7 = 0
7 + 0 = 7
0 + 7 = 7

House 2:
7 - 1 = 6
7 - 6 = 1
6 + 1 = 7
1 + 6 = 7

House 3:
7 - 2 = 5
7 - 5 = 2
5 + 2 = 7
2 + 5 = 7

House 4:
7 - 3 = 4
7 - 4 = 3
4 + 3 = 7
3 + 4 = 7

156.

7.9f. House 1:
8 - 0 = 8
8 - 8 = 0
8 + 0 = 8
0 + 8 = 8

House 2:
8 - 1 = 7
8 - 7 = 1
7 + 1 = 8
1 + 7 = 8

House 3:
8 - 2 = 6
8 - 6 = 2
6 + 2 = 8
2 + 6 = 8

House 4:
8 - 3 = 5
8 - 5 = 3
5 + 3 = 8
3 + 5 = 8

House 5:
8 - 4 = 4
4 + 4 = 8

157.

7.9g. House 1:
9 - 0 = 9
9 - 9 = 0
9 + 0 = 9
0 + 9 = 9

House 2:
9 - 1 = 8
9 - 8 = 1
8 + 1 = 9
1 + 8 = 9

(7.9g. continued on next page)

House 3:
9 - 2 = 7
9 - 7 = 2
7 + 2 = 9
2 + 7 = 9

House 4:
9 - 3 = 6
9 - 6 = 3
6 + 3 = 9
3 + 6 = 9

House 5:
9 - 4 = 5
9 - 5 = 4
5 + 4 = 9
4 + 5 = 9

158.
7.9h. House 1:
10 - 0 = 10
10 - 10 = 0
10 + 0 = 10
0 + 10 = 10

House 2:
10 - 1 = 9
10 - 9 = 1
9 + 1 = 10
1 + 9 = 10

House 3:
10 - 2 = 8
10 - 8 = 2
8 + 2 = 10
2 + 8 = 10

House 4:
10 - 3 = 7
10 - 7 = 3
7 + 3 = 10
3 + 7 = 10

House 5:
10 - 4 = 6
10 - 6 = 4
6 + 4 = 10
4 + 6 = 10

House 6:
10 - 5 = 5
5 + 5 = 10

159.
7.10a. 2, 0, 3, 3, 1

3, 0, 3, 4, 2, 3, 2, 4

4, 1, 4, 2, 4

1, 1, 4, 4, 3, 3, 0, 3

4, 3, 3, 2, 0

160.
7.10b. 0, 6, 5, 3, 6

1, 2, 5, 6, 4, 6, 5, 5

6, 1, 5, 1, 4

2, 0, 5, 6, 4, 6, 3, 5

6, 4, 5, 2, 3

161.
7.10c. 3, 6, 3, 5, 3

6, 1, 5, 3, 1

2, 2, 1, 5, 0

4, 4, 4, 1, 4

3, 2, 6, 2, 4

162.
7.10d. 4, 2, 8, 8, 1, 7, 6, 0

3, 8, 6, 7, 8

4, 7, 7, 8, 7, 7, 6, 8

6, 2, 7, 1, 0

5, 1, 7, 8, 5, 7, 3, 2

163.
7.10e. 6, 4, 10, 9, 3, 9, 8, 0

4, 10, 8, 9, 10

3, 9, 9, 10, 10, 10, 8, 10

6, 2, 9, 3, 1

7, 2, 9, 10, 6, 9, 1, 5

164.
7.10f. 6, 10, 6, 9, 7

10, 5, 9, 5, 3

1, 4, 2, 9, 5

9, 8, 8, 3, 8

7, 4, 10, 7, 8

166.
7.12a. 7, 3, 2, 5, 2

2, 4, 4, 2, 4

5, 5, 5, 6, 7

167.
7.12b. answers vary

168.
7.12c. 1, 0, 4

2, 5, 6

7, 3

CPSIA information can be obtained at www.ICGtesting.com
Printed in the USA
BVOW11s0548280415

397224BV00006BA/38/P